The Concept of Probability in Statistical Physics

Foundational issues in statistical mechanics and the more general question of how probability is to be understood in the context of physical theories are both areas that have been neglected by philosophers of physics. This book fills an important gap in the literature by providing the most systematic study to date of how to interpret probabilistic assertions in the context of statistical mechanics.

The book explores both subjectivist and objectivist accounts of probability, and takes full measure of recent work in the foundations of probability theory in statistical mechanics and in mathematical theory.

The book will be of particular interest to philosophers of science, physicists, and mathematicians interested in foundational issues, and also to historians of science.

Y. M. Guttmann is Associate Professor of Philosophy at Stanford University.

". . . essential reading for philosophers of physics interested in statistical mechanics and of deep interest to those in physics and mathematics interested in foundational issues."
                    Larry Sklar, author of *Physics and Chance*

". . . it will become the standard reference for philosophers of science on foundational questions of statistical physics and related matters. It will be likewise useful as an advanced textbook."
                    Jan von Plato, author of *Creating Modern Probability*

**Cambridge Studies in Probability,
Induction, and Decision Theory**
General Editor: Brian Skyrms

Advisory editors: Ernest W. Adams, Ken Binmore, Jeremy Butterfield,
Persi Diaconis, William L. Harper, John Harsanyi, Richard C. Jeffrey,
Wolfgang Spohn, Patrick Suppes, Amos Tversky, Sandy Zabell

The Concept of Probability in Statistical Physics

Y. M. Guttmann
Stanford University

CAMBRIDGE
UNIVERSITY PRESS

32 Avenue of the Americas, New York NY 10013-2473, USA

Cambridge University Press is part of the University of Cambridge.

It furthers the University's mission by disseminating knowledge in the pursuit of education, learning and research at the highest international levels of excellence.

www.cambridge.org
Information on this title: www.cambridge.org/9780521621281

First published 1999

A catalogue record for this publication is available from the British Library

Library of Congress Cataloguing in Publication data
Guttmann, Y. M. (Yair M.), 1957–
The concept of probability in statistical physics / Y. M. Guttmann.
p.   cm.
Includes index.
ISBN 0-521-62128-3 (hb)
1. Probabilities.   2. Statistical physics.   I. Title.
QC174.85.P76G88   1999   530.13 – dc21   98-40331   CIP

ISBN 978-0-521-62128-1 Hardback
ISBN 978-0-521-04217-8 Paperback

To the women in my life
my late mother Zipora, my wife Daniela, and my daughter Vita

Contents

Introduction

Probabilities have a puzzling character within the context of classical physics. Here, unlike in quantum mechanics, the use of probabilistic reasoning is not based on the existence of indeterminacy or objective lawlessness. Indeed, indeterminacy, in general, has no place in classical physics. In the very foundation of classical physics we find the assumption that, given the precise state of the world in one instance, the laws of physics determine its future states completely. Therefore, the introduction of probabilities into classical physics presents a problem. How are we to interpret statements from statistical physics whose abstract form is "The probability of A is p."? Do probabilistic statements form an integral part of the description of the physical world? Do they merely reflect our ignorance with respect to the precise state of the world? Can they be deduced from nonprobabilistic statements? Can they be understood in terms of frequencies? This book is dedicated to the study of these questions.

Readers who are not very familiar with statistical mechanics may want a few examples of physical statements using probabilities. The following is a list of statements, written in plain English, expressing physical truths (classified under the heading of thermodynamics) whose precise formulation requires probabilistic concepts:

1. Water freezes at zero centigrade.
2. When two bodies are in contact with one another, the heat flows from the warmer body to the colder one.
3. When you heat a container full of gas, the pressure on the walls of the container will increase.
4. When you add milk to your coffee and stir the cup, the two liquids will mix well. Further stirring will not separate the liquids.
5. A bomb explosion might turn a building into a heap of rubble; a

1

second explosion will not turn the heap of rubble back into a building.

6. The higher the temperature of a body, the hotter it feels to the senses.

The reason why the precise formulation of these statements involves probabilities is simple enough to be understood by a layman. We can construct special scenarios that are consistent with the laws of physics where each of the above statements will be false. These scenarios, physicists claim, are physically possible but extremely improbable; therefore, the above-mentioned statements are not absolute certainties but highly probable generalizations. This book is concerned with the precise meaning of such claims.

The most famous example of such a physically possible but extremely improbable scenario was given by Maxwell, the founder of statistical mechanics. The purpose of his example was to demonstrate that it is physically possible (although highly unlikely) that a cold body will transfer energy to a warmer body. This example was the following. Let A be an insulated container that is divided into two compartments, B and C. Both compartments are filled with gas, but the temperature of B is higher than the temperature of C. Between B and C there is a sliding door. When the door is open, molecules may flow from one compartment to the other. When it is closed, B and C are completely insulated. Imagine that near the door sits an "evil demon." When the demon sees a very "fast" molecule approaching from C toward the door, the demon opens the door and lets the molecule pass from C to B. When it sees a very "slow" molecule approaching from B toward the door, it lets it pass from B to C. Assuming that even "cold" bodies have some "fast" molecules and warm bodies have some "slow" ones, the demon will cause a heat transfer from the colder body to the warmer one! Suppose, next, that we have a second container that is an exact replica of the first. This time, though, there is no demon near the door. There is only a random device that opens the door at irregular intervals. Suppose that it so happens that the random device opened the door exactly when the demon did. In such a case, too, the same conclusion will follow. A heat flow will be created from C to B. Notice that the second case is completely consistent with the laws of physics and it contains no supernatural elements. It proves that, in very special circumstances, the "fast" molecules of a colder body

2

might be transferred to a warmer one, thereby contradicting the laws of thermodynamics.

Similar counterexamples may be constructed in the other cases as well. Take the fifth statement, for example. This statement is true by virtue of the second law of thermodynamics, namely, that the natural trend of things leads toward disorder. One of the consequences of the second law is that there is an inherent irreversibility in the physical world. This irreversibility is responsible for the fact that a bomb explosion will not transform a heap of rubble back into the original building. This irreversibility, though, is difficult to reconcile with the fact that the motion of the molecules that compose the rubble is completely reversible. To understand why the motion of the molecules is reversible, observe that classical physicists think of molecules as if they were billiard balls that obey a very simple law of motion, namely, that when the balls collide with each other or with the wall of a container, the angle of reflection is the same as the angle of incidence. This law is, of course, completely time-symmetric. Therefore, if we filmed the motion of the billiard balls and played the film backward, the fact that in the film the time flowed backward would not be detected by an observer (that is, until one of the balls stopped). However, if the reversed motion of the molecules obeys the same physical laws as the original motion, then it must be physically possible for the rubble molecules to be transformed back into their original state and create, again, a building! Indeed, Maxwell, who knew this argument, concluded that "The truth of the second law is, therefore, a statistical and not a mathematical truth, for it depends on the fact that the bodies we deal with consist of millions of molecules and that we never can get a hold of single molecules."[1]

Maxwell's answer demonstrates the importance of the discussion on the foundations of probability. As long as we do not have a good interpretation of our probabilistic concepts or at least a justification of the probabilistic arguments that appear in statistical mechanics, we have no way of understanding why certain types of events are considered improbable. As a result, we cannot fully understand statements such as the second law of thermodynamics and their consequences. Some writers believe that there are objective facts by virtue of which certain

---

[1] See Maxwell's review of Tait's "Thermodynamics," in his *Scientific Papers*, vol. II, p. 670.

3

events are extremely improbable. If this were the case, the second law would acquire a precise meaning. But can an objectivist conception of probability be reconciled with the deterministic character of classical physics? Other writers believe that probabilistic statements are merely subjective judgments. Does that mean, then, that we have no firm grounds for believing, for example, that water freezes at zero centigrade? These questions, and related ones, will occupy us for the rest of the book.

An informed reader may ask some further questions: "Why be preoccupied with classical mechanics? Was it not surpassed by quantum mechanics as the correct theory of the world?" The truth of the matter is that quantum statistical mechanics still owes much of its content and structure to its classical counterpart. When quantum phenomena become relevant, certain new parameters have to be introduced into the statistical mechanical framework. But the framework itself remains essentially the same. "But is it not the case that in quantum mechanics probabilities are introduced because of the underlying indeterminacy? Therefore, is it not the case that quantum probabilities are essentially different from the classical ones that you described?" The answer is that quantum statistical mechanics contains two types of probabilities that cannot be given a unified presentation. The first type of probabilities are related to the Schroedinger wave equation. Probabilities of the second type, on the other hand, that are used in quantum statistical mechanics are defined in exactly the same way as they are defined in the classical theory.

Let us now sketch the general structure of the book.

We begin the first chapter with a survey of the early history of statistical mechanics. The survey introduces the more elementary aspects of the work of Maxwell, Boltzmann, and Gibbs. These writers labored to integrate probabilistic reasoning within physics. The result was a physical theory with an incredible predictive power. But the marriage between the probabilistic and the physical concepts was not an easy one. Even today, almost 150 years after the beginning of statistical mechanics, the basic methodological issues are not completely resolved. A question that may be raised, then, is whether we can retain the predictions of statistical mechanics (henceforth SM) without introducing physical probabilities into the picture. This question gives rise to an eliminative program that we shall label ultrasubjectivism. The object of this program is to strip down SM and present it in the form of a general probabilistic inference based solely on a priori principles of inductive reasoning and empirical data. (One advantage of presenting SM in this manner is that the empirical predictions that SM makes become independent of any assumption about the underlying determinism. This makes it easier to unify classical and quantum SM.) The originator of the ultrasubjectivist program is the physicist E. T. Jaynes. In the first chapter we review Jaynes's ideas and argue that his ultrasubjectivist program fails to yield a general and defensible version of SM.

The second chapter begins with a historical survey of statistical mechanics after Gibbs. The survey emphasizes one modern tendency, namely, the ergodic approach to SM. This approach gave rise to some of the most interesting and important developments in statistical mechanics, but it is rather difficult to present it as a coherent foundational approach. One of the ways of understanding the ergodic approach is as an approach to SM that aims mainly to develop arguments proving the consistency of thermodynamics and classical mechanics. This approach is consistent with a more moderate subjectivist approach than the one we present in the first chapter. The adherents of this more moderate variant of subjectivism attempt to reconcile the "frequentist" aspect of the ergodic approach to SM with subjectivism. This approach, we argue, does not do justice to statistical mechanics. Some of the research conducted in this century yielded very impressive results that go far beyond the search for inductive consistency proofs.

The third chapter is dedicated to the question of whether the probabilities we find in SM can be eliminated. That is, we shall ask ourselves whether SM probabilities can be reduced to other concepts. We shall discover that in some rarefied sense, the construction of Haar measures can be seen in this light. However, in most cases of physical interest a strict reduction of the probabilities is impossible.

In the fourth chapter we introduce into the discussion some topological notions. These notions will allow us to formulate the idea that stochastic behavior is the result of instability, that is, of the fact that small undetectable differences in the initial conditions may result in macroscopically observable differences. The important notion of instability, though, cannot replace all of the necessary probabilistic concepts.

Finally, in the fifth chapter we discuss three solutions to the puzzles that occupied us throughout the book. We outline three ways of thinking about statistical mechanical probabilities. The first way is a new version of the subjectivist approach, and the second is a new way of formulating the ergodic approach to statistical mechanics. Finally, we shall return to the writings of Gibbs and argue that his view, which we shall call a pragmatist approach to statistical mechanics, constitutes a highly satisfactory solution.

It might be worth mentioning that books on statistical mechanics are divided into two groups: those that are centered on equilibrium statistical mechanics and those that are more engaged in the study of the nonequilibrium case. The present book belongs to the former group; it is somewhat biased toward the equilibrium case. The reason for the preference is that we believe that, for the purposes of studying the nature of statistical mechanical probabilities, equilibrium statistical mechanics offers a better conceptual framework. If, on the other hand, one is interested in studying the concept of irreversibility, one must study nonequilibrium statistical mechanics. Because this book is centered on the foundational study of probabilistic concepts, we shall study closely the case of equilibrium statistical mechanics.

The book can be read in four different ways. First, it can be read as an essay in theoretical physics on the different ways to formulate SM. Of particular importance are the different formulations of the second law of thermodynamics. In each chapter we introduce a new type of formulation of SM that will be given a philosophical motivation. These formulations then are tested for appropriateness and adequacy. In particular, we review in some detail the contributions of the ergodic theory and of topological dynamics to the foundations of SM.

Second, we follow the historical developments that led to the modern formulation of SM. We begin with the work of the "founding fathers," Maxwell, Boltzmann, and Gibbs. In addition, we discuss the work of Poincaré, the Ehrenfests, Lyapunov, von Neumann, Birkhoff, Hopf, Khinchin, Kolmogorov, Abraham, Sinai, and others on the different ways to characterize order and disorder through the concepts of quasi-periodicity, recurrence, stability, ergodicity, and entropy. These authors and concepts are placed in a historical context that offers the reader a taste of the history of the modern science of mechanics.

Third, and most importantly, the book is meant to be a contribution to the study of the foundations of probability theory. From this perspective, SM is treated as a context in which the different foundational standpoints can be given fair and precise formulations. In particular, SM presents an important challenge to the subjectivists.

Finally, we believe that the philosophical study of SM yields many new insights that contribute to the discussion on various philosophical topics. Among these topics are the discussions on the concept of scientific reduction, correspondence principles, and other relations holding among scientific theories; the relations between mathematics and physics; and the role of mechanical models in physics and the interpretation of theoretical entities. Other topics that appear in the discussion include the concept of supervenient properties, the nature of induction, the "emergence" of properties, and the nature of physical modalities. A topic that is discussed frequently is the nature of de Finetti's subjectivist methodology. A more general philosophical thesis that emerges is the relevance of pragmatist philosophy to the discussions on the philosophy of physics.

The general mathematical level of the presentation is relatively high, and we regret this fact. We feel, however, that a precise and faithful formulation of the ideas of SM would be beneficial to a deeper understanding of the material. The literature on the philosophy of science is full of examples that prove that the attempts to "predigest" science for the benefit of an "uninformed philosophical reader" are usually quite unsuccessful. At their best, nontechnical presentations leave some of the most important ideas out of the discussion. In many cases, though, such presentations contain serious mistakes that are transmitted, for years and years, and affect the philosophical discussion on other unrelated issues. The literature on the subject of scientific reduction is a case in point. One finds there the view that SM furnishes a prime example of the reduction of scientific theories. Even our very elementary introduction, though, is sufficient to demonstrate that this cannot be the case. To reduce a theoretical statement means that one may replace it with a synonymous statement from another theory. The presence of statistical concepts in the arguments of SM, however, makes such replacements impossible. To achieve a reduction, we are always instructed to ignore some improbable but physically possible events. Therefore, even at their best, SM considerations cannot help us to effect a strict reduction.[2] The effects of this mistaken view are not confined to the discussion on the philosophy of physics. Many philosophers believe that the example of SM lends some initial plausibility to

---

[2] This does not mean that the notion of reduction cannot be suitably generalized. In fact, we shall discuss many such generalizations throughout the book. However, the relations between thermodynamics and mechanics demonstrate that those cases where reduction in physics led to fruitful results have very little to do with the dogmatic physicalism of most philosophical reductionists.

a more general reductionist attitude. This reductionist attitude in turn makes it even harder to understand the character of SM correctly. The acceptability of SM, with its statistical assumptions, is predicated on its *utility* and not on the fact that it takes us closer to the "real furniture of the world."[3] Other examples of the abuse of the discussion on statistical mechanics abound in the literature. We find them, for example, in connection with the discussion on the role of theoretical models, on the interpretation of theoretical entities, and in the discussion on the relation between phenomenological and fundamental laws in physics. We shall return to the discussion of these issues later in the book.

We would like to emphasize, though, that in writing this book our primary aim was not to defend science from nosy uninformed intruders. On the contrary, this book is motivated by the wish to democratize science, make the connections between science and other areas of culture more evident, and make scientific ideas available to intelligent nonspecialists without unduly popularizing them. Our belief is that one should be encouraged to develop an informed critical relation to science rather than blindly follow the word of the experts. This belief may justify, at least in part, the audacity inherent in the attempt by a philosopher to convey, within a single book, some aspects of a large and complex literature to an audience of philosophers.

The last point provides us with an opportunity to anticipate and answer a question that some readers may wish to ask: Is it the task of a philosopher, rather than the job of a physicist or a mathematician, to write a book on the nature of probabilities in statistical mechanics? The answer is that a philosophical temperament is, indeed, required of those who attempt to go to the bottom of the subject. For most physicists, the topic is too "academic," because discussing it is not likely to improve our ability to make better predictions, to discover new effects, or to explain phenomena that we do not yet understand. Besides, most physicists believe that Gibbs's statistical mechanics is fine just the way it is, and, in a sense, we believe that they are right. Mathematicians, on the other hand, usually shy away from discussing mathematical concepts in nonmathematical terms. They believe that every meaningful question about mathematical objects, about probability functions, for example, should be formulated as a mathematical question. Therefore,

[3] See, for example, the discussion on Ising models in Chapter 2. Such models, which furnish prime examples of reduction in physics, are extremely unrealistic.

when mathematicians encounter a question about a mathematical object that is not strictly speaking a mathematical question, they suspect that the question is not a meaningful one. They, too, are generally right. However, if neither the physicist nor the mathematician is prepared to answer our questions and tell us whether the probabilistic statements of statistical mechanics express objective truths or merely subjective opinions, who is going to do so? It would seem that this difficult assignment is allotted to the philosophers of physics. Indeed, in the last 20 years or so several books and articles have been written about the philosophy of statistical mechanics. This book is an attempt to contribute to this recent tradition.

After much deliberation we decided to treat every chapter as an autonomous unit. The disadvantage is a certain amount of unavoidable repetitiveness in the exposition of the elementary concepts. We did so to make it possible to read each of the chapters separately. To the main body of the book we added three appendices, one that includes a survey of some of the mathematical concepts that appear in the book and another that contains a survey of the discussion on the foundations of probability. We recommend to readers who do not have the necessary background in these areas to read these appendices as introductory chapters.

1

The Neo-Laplacian Approach to Statistical Mechanics

1.1. THE INTRODUCTION OF STATISTICS INTO PHYSICS

In 1859, J. C. Maxwell presented his "Illustrations of the Theory of Gases" to the British Society for the Advancement of Science.[1] In the paper he applied the theory of errors to the problem of finding "the average number of particles whose velocities lie between given limits, after a great number of collisions among a great number of particles."[2] The conclusion that Maxwell reached was that "the velocities are distributed among the particles according to the same law as the errors are distributed in the theory of the method of least squares."[3] In the second part of the paper, Maxwell distinguishes between the motion of translation of the system as a whole and the motion of agitation. The latter is characterized as a state where "the collisions are so frequent that the law of distribution of molecular velocities, if disturbed in any way, will be reestablished in an appreciably short time."[4] This last statement may serve to explain why "Illustrations" is considered to be the beginning of modern statistical mechanics.[5]

The discussions on the theory of errors itself started more than 100 years earlier. Simpson discussed the idea of averaging the results of various observations in 1757. Bernoulli introduced the assumption that large errors should be regarded as less probable than small errors in

---

[1] "Illustrations of the Dynamical Theory of Gases," *Phil. Mag.* 4(19) (1860). Reprinted in *The Scientific Papers of James Clerk Maxwell*, ed. W. D. Niven, Dover (New York, 1963).
[2] Ibid, p. 380.   [3] Ibid, p. 382.   [4] Ibid, p. 392.
[5] Another remarkable result that is communicated in the paper is the fact that the viscosity or "internal friction," as Maxwell called it, is independent of the density of the gas and is proportional to the square root of the absolute temperature. Ibid, p. 391.

1777, and Lagrange introduced the least-squares method in an undated manuscript written in the 1770s.[6] Other important contributors to the early stages of the theory of errors are Gauss and Legendre, but it was Laplace who systematized the theory and showed how to apply it in various contexts. Of particular interest were the applications of the theory of errors to astronomy and economics.

Laplace regarded the theory of probability in general, and the theory of observational errors in particular, as complementary theories to mechanics. Mechanics, for Laplace, provided an objective description of the way bodies move. The theory of errors, on the other hand, is a heuristic that we use to derive useful predictions even from observations that contain errors. Given this background we can see at once that Maxwell's application of the theory of errors contained a novel character. It introduced probabilities into the science of mechanics itself. In other words, probabilities became part and parcel of a science whose aim was to provide an objective description of the properties of systems in motion. This idea became the cornerstone of statistical physics.

According to Gillespie,[7] Maxwell arrived at his derivation after reading an article by Sir John Herschel in which he reviewed the theories of Quetelet. Herschel believed that the theory of errors was important enough to be understood and applied even by nonspecialists. Hence he attempted to provide a particularly simple and illuminating proof of the law of errors that proceeded from the following three postulates:[8]

(1). The probability of a compound event, or of the occurrence of two or more simple events is the product of the probabilities of its constituents considered singly.
(2). That there exists a relation or a numerical law of connexion (at present unknown,) between the amount of error committed in any numerical deter-

---

[6] For a historical survey, see I. Todhunter's *History of the Theory of Probability*, Macmillan (Cambridge, UK, 1865).
[7] See Gillespie in *Scientific Change*, ed. A. C. Crombie, Basic Books (New York, 1963), p. 431. For a historical study of the development of the statistical point of view and its applications, first to sociology and then to physics, see J. T. Merz's *History of European Scientific Thought in the 19th Century*, Peter Smith (Massachusetts, 1896), ch. 12.
[8] See Stephen G. Brush, *The Kind of Motion We Call Heat*, North Holland Publishing (Dordrecht, 1976), p. 184.

mination and the probability of committing it such that the greater the error, the less its probability.

(3). That the errors are equally probable if equal in numerical amount, whether in excess or in defect of the truth. This latter postulate necessitates our assuming the function of probability to be what is called in mathematical language an even function or a function of the square of the error. The form of the function which is asserted to be generally valid may be determined by considering a special case, namely, a ball dropped from a given height with the intention that it falls on a given mark. Fall as it may, its deviation from the mark is error and the probability of that error is the unknown function of its square i.e. the sum of the squares of its deviations in any two rectangular directions. Now, with the probability of any deviation depending solely on its magnitude, and not on its direction, it follows that the probability of each of the rectangular deviations must be the same function of its square. And since the observed oblique deviation is equivalent to the two rectangular ones, supposed concurrent and which are essentially independent from one another, and is, therefore a compound event of which they are the simple independent constituents, therefore its probability will be the product of its two separate probabilities. Thus the form of our unknown function comes to be determined from this condition, viz. that the product of such a function of two independent elements is equal to the same function of their sum. But it is shown in any work on algebra that this property is peculiar characteristic of, and it belongs only to, the exponential or logarithmic function. This then, is the function of the probability of committing this error.

What is interesting about Herschel's example is that the deviations from the mark can be regarded as errors, but they can also be regarded as the outcomes of a process that contains a stochastic element. In fact, Herschel's presentation indicates that the theory of errors may be applied even in cases where the notion of an error does not seem to arise at all! That was precisely the way Maxwell read Herschel's review.[9]

---

[9] According to Gillespie, this feature of Herschel's example was fully intentional. Herschel was following Quetelet's aspiration to show that the law of errors had a wider range of application than previously suspected. In one of his studies, Quetelet measured the perimeter of the chest of Belgian soldiers. He found that the mean value was slightly over 40 inches. This value was also the most probable value. Quetelet argued that the distribution of the variations around the mean value was so similar to the distribution of errors that we may regard the variations around the mean as if they were God's errors. See also von Plato's *Creating Modern Probability*, Cambridge Univ. Press (Cambridge, UK, 1994), pp. 73–74.

Maxwell was not preoccupied with the kinetic theory of Clausius before 1859. (In fact, in a letter to Stokes, he indicated his feeling that the kinetic theory of Clausius and Herapath could be refuted.[10]) However, Maxwell was acquainted with the works of Kroenig and Clausius on the subject, which they published, respectively, in 1856 and 1857. So it is tempting to conjecture that, after reading Herschel's article, Maxwell saw how some of Herschel's ideas could shed an unexpected light on the peculiarities of the kinetic theory of gases. Recall that Kroenig and Clausius tried to calculate the pressure that the molecules of a gas exert on the wall of a container through their "heat motion." This calculation seemed enormously complicated, so, as it often happens in physics, Kroenig gave up the idea of describing the dynamics of the molecules precisely and, instead, constructed a very simple model of molecular motion that allowed him to calculate the pressure. According to this model, the gas molecules move with the same velocity along lines that are parallel to the three rectangular walls of the container (one-third along each axis). This model proved rather useful; it allowed Kroenig to derive the ideal gas law.[11] However, what could possibly justify the above-mentioned oversimplifying assumptions? Kroenig's justification was rather interesting. Anticipating ideas that Poincaré developed more than a half-century later,[12] he explained that "the path of each molecule must be so irregular that it will defy all calculations. However, according to the laws of probability theory, we can assume a completely regular motion in place of this completely irregular motion."[13] This formulation is certainly obscure, but behind it there is an indication of the existence of a principle according to which there are circumstances when one may legitimately select a simple

[10] For a discussion on the early work of Maxwell, see Stephen G. Brush, *The Kind of Motion We Call Heat*, North-Holland Publishing (Amsterdam, 1976), p. 189. Also see Brush's *Statistical Physics and the Atomic Theory of Matter*, Princeton Univ. Press (Princeton, 1985), pp. 56–68 and C. Truesdell, "Early Kinetic Theory of Gases," *Arch. Hist. Ex. Sci.* 15 (1975).

[11] See Brush's *Statistical Physics and the Atomic Theory of Matter*, pp. 46–54.

[12] See Poincaré's "Chance" in *Science and Method*, Dover (New York, 1957), pp. 64–90. See also "The Milky Way and the Theory of Gases," ibid, pp. 253–269, where he writes: "In face of a semidisorder we should be forced to despair. But in extreme disorder (this) statistical law reestablishes a kind of average or mean order in which the mind can find itself again. It is the study of this mean that constitutes the kinetic theory of gases" (p. 257).

[13] See P. and T. Ehrenfest, *The Conceptual Foundations of the Statistical Approach in Mechanics*, Cornell (Ithaca, NY, 1959), p. 4.

"average" case from the set of all possible cases, and base one's calculation on this simple case. This principle is the law of errors, at least the way Maxwell understood it.

Maxwell wrote:

Let N be the whole number of the particles. Let x,y,z be the components of the velocity of each particle in three rectangular directions, and let the number of particles for which x lies between x and $x + dx$ be $Nf(x)dx$, where $f(x)$ is a function of x to be determined.

The number of particles for which y lies between y and $y + dy$ will be $Nf(y)dy$; and the number for which z lies between z and $z + dz$ will be $Nf(z)dz$, where f always stands for the same function.

Now, the existence of the velocity x does not in any way affect that of the velocities y or z, because these are all at right angles to each other and independent, so that the number of particles whose velocity lies between x and $x + dx$, between y and $y + dy$, and between z and $z + dz$ is

$$Nf(x)f(y)f(z)dxdydz.$$

If we suppose the N particles to start from the origin at the same instance, then this number will be the number in the element of volume ($dxdydz$) after a unit of time, and the number referred to as the unit of volume will be

$$Nf(x)f(y)f(z).$$

But the directions of the coordinates are perfectly arbitrary, and therefore this number must depend on the distance from the origin alone, that is,

$$f(x)f(y)f(z) = F(x^2 + y^2 + z^2).$$

Solving this functional equation, we find

$$f(x) = Ce^{Ax^2}, \qquad F(r) = C^3 e^{Ar^2}.$$

If we make A positive, the number of particles will increase with the velocity, and we should find the whole number of particles infinite. We therefore make A negative and equal to $-1/\alpha^2$, so that the number between x and $x + dx$ is

$$NCe^{-x^2/\alpha^2}dx.$$

Integrating from plus to minus infinity, we find the whole number of particles,

$$NCa\sqrt{p} = N$$

so that

14

$$C = 1/\alpha\sqrt{p}$$

and $f(x)$ is therefore

$$1/\alpha\sqrt{p}\,e^{-x^2/\alpha^2}.$$

Finally, by using this formula, Maxwell concluded that the mean velocity of the particles is $2\alpha/\sqrt{p}$ and the mean square velocity is $3\alpha^2/2$.[14]

1.2. MAXWELL'S SECOND DERIVATION OF THE VELOCITY DISTRIBUTION

The assumption that v_x, v_y, and v_z, the three velocity components of a single molecule, are independent of one another was essential to the way in which Maxwell derived the velocity distribution. By using this assumption, he made Herschel's method applicable to the theory of gases. Nevertheless, Maxwell was not altogether satisfied with his derivation. In fact, he regarded the assumption to which we alluded as "precarious".[15] Indeed, why should we regard the three velocity components, which are which are mathematical constructs without a separate physical existence, as separate independent events? As Maxwell put it, there is no reason why the knowledge of v_x should not affect our assumption about the probabilities of v_y and v_z. Therefore, in a second article dated 1866 and entitled "The Dynamic Theory of Gases," Maxwell attempted a second derivation of the velocity distribution.

The starting point of "The Dynamic Theory of Gases" is an observation that is at once immediate and extremely important. So far, the adherents of the kinetic theory of gases explained the existence of pressure in terms of kinetic motion. There is, however, a second phenomenon to be explained, namely, the uniformity of pressure that is exerted on the different walls of the vessel.[16] Maxwell observed that there must be a mechanism that is responsible for the lack of observable

---

[14] Note that, unlike the error curve, the Maxwell distribution function $P(v) = Cv^2 e^{-v^2/\alpha^2}$ is not symmetric (because of the additional dependence on the square of the velocity). See Brush's *Statistical Physic and the Atomic Theory of Matter*, Princeton, p. 60.

[15] See Maxwell's "On the Dynamic Theory of Gases," *Phil. Trans.* 157(49) (1867), p. 43.

[16] Ibid, p. 39.

fluctuation in the pressure in different directions. The mechanism is the collisions between the molecules. (Maxwell called it the mutual action of the molecules.) Through their collisions, the molecules can "communicate" motion.[17] Such communication is possible even between molecules of different types.[18] The collisions between the molecules will, in general, change the distribution of the positions and velocities among the molecules. Therefore, for an equilibrium distribution to be possible, a special relation has to hold between the collisions and the equilibrium distribution.

In "The Dynamic Theory of Gases," Maxwell attempted to derive the equilibrium distribution from the following two assumptions:

1. The velocities of the colliding molecules are uncorrelated. Hence, if $P(v_1, v_2)$ is the probability that a pair of colliding molecules will have, respectively, the velocities v_1 and v_2, then

$$P(v_1, v_2) = P(v_1)P(v_2),$$

where $P(v_1)$, $P(v_2)$ are the probabilities of choosing at random a molecule with the respective velocities.

2. Let v_1, v_2 be the velocities of a pair of colliding molecules before the collision and v'_1, v'_2 their velocities after the collision. Maxwell's second assumption was that the probability that a colliding pair will undergo the velocity change $v_1, v_2 \rightarrow v'_1, v'_2$ because of the collision will be the same as the probability of the velocity change $v'_1, v'_2 \rightarrow v_1, v_2$ That is, the "reverse" motion is as probable as the original motion.[19]

From the two assumptions it follows that

$$P(v_1, v_2) = P(v'_1, v'_1)$$

or

$$P(v_1)P(v_2) = P(v'_1)P(v'_2).$$

On the other hand v_1, v_2, v'_1, v'_2 are also mutually constrained by the energy conservation law. That is:

[17] See Maxwell's "On the Dynamic Theory of Gases," p. 3.   [18] Ibid, p. 30.
[19] In quantum mechanics, this property is reflected in the fact that the matrix of the transition probabilities is symmetric.

$$\frac{1}{2}m_1v_1^2 + \frac{1}{2}m_2v_2v_2^2 = \frac{1}{2}m_1v'_1^2 + \frac{1}{2}m_2v'_2^2.$$

From these Maxwell derived the equation

$$P(v_1) = (N/\alpha^3 p^{3/2})e^{-(v_2/\alpha^2)},$$

which completes the derivation.

Note that the second derivation brings into the fore new mechanical aspects of the kinetic theory. This is already evident in the facts that the collisions between the molecules are investigated and use is made of the conservation of energy. The article itself contains many more interesting physical investigations, among them a derivation of the relation between the density of a gas and its weight, the "Law of equivalent volumes," as Maxwell called it. In another important passage, Maxwell explained the phenomenon of diffusion. As we remarked earlier, the realization that the pressure is the same, irrespective of direction, was an important insight.

What, though, is the status of Maxwell's two assumptions? It seems that no justification was offered. Behind the assumptions is an idea that is at the same time compelling and problematic, namely, that the positions and velocities can be assumed to be randomized. Moreover, it is assumed that, because of the collisions, the molecules will remain randomized. These assumptions do not offend common sense. So Maxwell clearly progressed since the first derivation. But the question remains: Why are we justified in making these assumptions?

The next major step in the development of the kinetic theory of gases was Boltzmann's 1868 paper.[20] In this paper, Boltzmann managed to derive the velocity distribution under less idealized conditions, namely, external forces were not assumed absent. Boltzmann considered, again, the case of a pair of colliding molecules. This time, though, one of the molecules was acted on by a force whose potential is V. Because of the existence of forces, the condition of the conservation of energy becomes

$$\frac{1}{2}m_1v_1^2 + V(x_1) + \frac{1}{2}m_2v_2^2 = \frac{1}{2}m_1v'_1^2 + V(x'_2) + \frac{1}{2}m_2v'_2^2.$$

---

[20] See "Studien ueber das Gleichgewicht der lebendigen Kraft zwischen bewegten materiellen Punkten," *Sitzungberichte, K Akademie der Wissenschaften, Wien, Mathematisch-Naturwissenschaftliche Klasse* 58 (1868), pp. 517–560.

Boltzmann proved that

$$f(v) = c[-h(1/2 mv^2) + V(x)],$$

where both c and h are constants. The constant h is related to the absolute temperature T. In fact, it is now common to set

$$h = 1/kT,$$

and k is referred to as Boltzmann's constant. Once external forces are introduced, the resulting velocity distribution is called the Maxwell–Boltzmann distribution. The physical meaning of the Maxwell–Boltzmann distribution law is that it is highly probable that

$$\frac{1}{2}mv^2 + V(x) = kT.$$

It is important to mention that Boltzmann's derivation started from the same assumptions that Maxwell made in "The Dynamic Theory of Gases." In his later writings, he became increasingly aware of the problematic status of the assumptions to which he introduced the term *Stosszahlansatz*, or the assumption of molecular chaos. This is evident, for example, in Boltzmann's careful exposition of Maxwell's ideas in *Lectures on Gas Theory*.[21] Throughout his life Boltzmann tried to find a purely mechanical justification to the Stosszahlansatz.

1.3. BOLTZMANN'S ATOMISM AND ITS CRITICS

In the same paper of 1868, Boltzmann presented a third derivation of the equilibrium velocity distribution, that was rather different from the earlier ones. Boltzmann assumed that the gas molecules had a fixed amount of energy and looked for an expression for the number of different ways of dividing this amount between the different molecules. This method allowed Boltzmann to derive Maxwell's velocity distribution in a purely combinatorial way. Another important aspect of this new derivation was that the assumptions on the nature of the collisions were no longer necessary. They were replaced by a single purely probabilistic assumption, that is, Boltzmann assumed that all of the different possibilities were equally probable. This assumption

---

[21] See Boltzmann's *Lectures on Gas Theory*, Engl. trans. by S. G. Brush, Univ. of California Press (Berkeley, 1964), pp. 35–49. Originally published in 1896.

was sufficient for a mathematical proof that the most probable distribution of energy corresponds to the Maxwell–Boltzmann distribution.[22]

We shall return to the details of the derivation later in this chapter. However, it is important to emphasize that, even after his combinatorial derivation, Boltzmann kept looking for a more mechanical version of the kinetic theory of gases. Boltzmann's models were atomistic. His starting point was that gases are composed of a large number of molecules whose motion is governed by mechanical laws. He believed that the construction of such models for gas systems, even when they involve ad hoc assumptions, was an indispensible tool for gaining more detailed and accurate physical knowledge. Indeed, even the most scant attention to the development of statistical mechanics (SM) confirms this point. All of Maxwell's ideas, for example, sprung directly from his molecular model of matter. Boltzmann, who continued the work of Maxwell, saw this point clearly and defended the use of the molecular models on the basis of the productive ideas to which it gave rise. Without such constructions, Boltzmann contended, Maxwell and he would not have been able to even formulate educated guesses concerning the properties of pressure or the mechanism of heat transfer.[23]

For this form of atomism Boltzmann was criticized rather severely by a group of physicists and mathematicians. The point of contention was a methodological one. Boltzmann's detractors believed that science should not depend on special speculative assumptions or on fanciful constructions of models. They believed that scientific theories should be divisible into two parts, a purely mathematical one that expresses the functional dependencies between various parameters and a purely empirical part, where the actual values of the parameters

---

[22] The replacement of kinetic assumptions with probabilistic ones marks, according to some, the beginning of statistical mechanics proper.

[23] This point is not easily appreciated and is often misconstrued. On the surface it is concerned merely with the psychology of inquiry. But in fact one should appreciate the fact that, to apply the tools of mathematical statistics, one needs to assume that the phenomenon in question is the result of a multitude of small uncorrelated causes, each of which is negligible or nearly so. This assumption, of course, does not amount to full-blown atomism. Gibbs saw this point clealy, and, as we shall see shortly, he viewed SM as furnishing an analogy that does not carry with it ontological commitments. But Gibbs's point of view gives rise to complaints of the opposite kind: One cannot help but feel that Gibbs's theory suffers from its use of the term analogy in terms of explanatory power.

are measured. From this perspective Boltzmann's mechanical models had no place or purpose in physics.

Another group of critical assessments of Boltzmann's kinetic theory was advanced by writers who were concerned with the issue of irreversibility.[24] Loschmidt, Zermelo, and Poincaré were the chief protagonists of this line of criticism.[25] These writers charged the kinetic theory with an inconsistency: the microscopic processes it postulates are taken to be reversible. Howver, the macroscopic phenomena that these processes give rise to are irreversible. Because we shall return to this issue in the next section, we will not discuss this point at length. However, the reader should appreciate the fact that, if the kinetic theory of gases was considered completely victorious in the 1870s, it was almost universally rejected by 1900. The critics of Boltzmann, especially Mach, Duhem, Poincaré, Kirchhoff, Helm, and Ostwald, were successful in portraying Boltzmann's mechanistic attitude as an unsophisticated and uncritical scientific realism. This critique became the philosophical basis for the views discussed a few decades later in the meetings of the Vienna circle.

Around 1904 began a period of positive reassessment of the kinetic theory of gases. The particular timing of this development had to do with the fact that some of Boltzmann's methods found applications in Planck's early studies of quantum phenomena.[26] Another important reason was Einstein's theory of Brownian motion, which gave Boltzmann's theory new empirical support. An even earlier renewed

---

[24] A third problem was the issue of specific heats. In essence, the problem is the following. One of the consequences of the equipartition theorem is that the ratio of specific heats at constant pressure to that in constant volume should be given by the expression $n + 2/n$, where n is the number of degrees of freedom of a gas molecule. In reality, however, the number of degrees of freedom seems to be smaller. This disappearance of degrees of freedom could have been explained if we had thought of a gas molecule as composed of rigidly connected material points. But this picture contradicts the fact that atoms were proven to have a rich internal structure. Many spectral lines were observed, each corresponding to a degree of freedom of the atom. The existence of many degrees of freedom implies the fact that the ratio $n + 2/n$ should have been close to unity. This, however, is not true for the known types of gases. This issue was resolved only with the advent of quantum theory. For a book on the subject, see T. Kuhn, *Black Body Theory and the Quantum Discontinuity 1894–1912*, Claredon (Oxford, UK, 1978).

[25] We will continue the discussion of these issues in Chapter 2. For an extensive discussion on the argument between Boltzmann and Zermelo see von Plato, *Creating Modern Probability*, ibid, pp. 89–92.

[26] See Brush, *Statistical Physics*, ibid, pp. 105–114.

support for Boltzmann's theories came after Gibbs's 1902 "Elementary Principles in Statistical Mechanics." Gibbs managed to present the investigations of Maxwell and Boltzmann in a favorable light that drew sympathy even from some of Boltzmann's archenemies.[27] At first blush, this might seem strange. Gibbs's line of thinking clearly continued that of Maxwell and Boltzmann, and Gibbs did not try to conceal this fact. One should remember, though, that the controversy surrounding Boltzmann's view was not always centered around his defense of atomism per se. By 1900, Boltzmann's writings comprised nearly 2000 pages of difficult and seemingly unsystematic treatments of various aspects of gas theory. Consequently, he was attacked for being unclear and uneconomical (Mach's term[28]), and for constructing theories that included too many special assumptions. In this regard, Gibbs's book was completely different. It was concise and mathematically informed, and paid special attention to the differences between the abstract theory and its applications. These qualities were probably responsible for the immediate popularity of Gibbs's book in Europe.

1.4. ON GIBBS'S CONCEPTION OF SM AND SOME ATTEMPTS TO REVISE IT

Gibbs's book begins with a review that takes us back to the very origins of statistical mechanics. (Incidentally, the term statistical mechanics was coined by Gibbs.) He once more confronts the question of the application of the theory of errors to mechanics. Gibbs reminds us that, when we calculate the value of any variable that depends on the position and momenta of the particles of a system with many degrees of freedom, there will be a certain interval of values $[A - a, A + a]$ such that it is vastly probable that the system will assume values in this inter-

---

[27] Ostwald translated "On A Heterogeneous System of Gases," an early paper of Gibbs's into German, and Zermelo wrote the introduction to the German translation of Gibbs's *Elementary Principles in Statistical Mechanics*, Yale Univ. Press (New Haven, CT, 1902).

[28] Mach wrote: "The atom must remain a tool for representing phenomena, like functions of mathematics. Gradually, however, as the intellect, by contact with its subject matter, grows in discipline, physical science will give up on its mosaic play in stones and will seek out the boundaries and forms of the bed in which the living stream of phenomena flows. The goal which it has set itself is the simplest and the *most economical* abstract expression of facts." (The italics are in the original text.) See Mach's *Popular Scientific Lectures*, Eng. trans. T. J. McCormick, Open Court Classics (LaSalle, IL, 1985). Originally published in 1894.

val. Moreover, very often the most probable value is the mean value. Gibbs emphasized the abstract character of this point and presented it as an outcome of the theory of errors. More precisely, Gibbs employed the theory of errors to simplify the computations of the values of various dynamic parameters. The simplification process consisted of a demonstration that, although in reality a variable may have infinitely many values, we may behave as if the values were identical to the mean value. What allows us to affect such a reduction is the theory of errors.[29] From a philosophical point of view, Gibbs's argument is interesting. As we shall see shortly, the theory of errors provides a rather rigorous defense for the idea that a useful fiction should be preferred to more realistic but useless ways of modeling the phenomena.

The more precise setting of the problem required Gibbs to define the notion of an "ensemble" (a notion introduced by Boltzmann, who called it "Ergode"). An *ensemble* is a collection of systems of a similar kind that differ from one another in the particular values which their parameters assume. The "relative proportion" of systems belonging to the ensemble whose parameters lie within the various intervals determines a density function. A normalized density function is a probability function. It assigns numerical probabilities to the event that a randomly chosen system will assume a particular value that is located within certain intervals for a given parameter. (We shall return to these issues later in the chapter.) As you may recall, Maxwell calculated the velocity distribution for systems with a fixed number of particles and a fixed energy. Boltzmann proved that the same type of distribution is obtained in the presence of forces. Gibbs continued the research by addressing two additional idealizing assumptions. Gibbs argued that (1) no real system is completely isolated, and hence we cannot expect systems to have a completely fixed energy; and (2) in many contexts the assumption that the number of particles is fixed can be limiting. (Chemical reactions may reduce or increase the number of degrees of freedom.) His goal, then, was to obtain the equilibrium distribution even when these considerations were taken into account. Gibbs's first step was to consider an ensemble of systems whose energy is allowed to vary. The probability that the system will have the energy level E was taken to be proportional to e^{-bE}, where $b = 1/kT$. Such an ensemble was called by Gibbs a *canonical ensemble*. Gibbs proved that, as

---

[29] Gibbs maintained that a necessary assumption is the constancy in time of the probabilities.

the number of particles grows to infinity, the canonical distribution coincides with the Maxwell–Boltzmann distribution. (Gibbs introduced the term *microcanonical distribution* for the Maxwell–Boltzmann distribution.) Therefore, one may regard systems that are in equilibrium with their surroundings as if they were isolated systems.[30]

Gibbs next considered an ensemble of systems with a varying number of particles. The number of systems having N particles was assumed to be proportional to e^{-mN} (m is the chemical potential). Such ensembles are known as *grand canonical ensembles*. Again, when the number of particles approaches infinity, the grand canonical distribution coincides with the microcanonical one. In practice, it is often much easier to use the canonical and grand canonical distributions than to calculate the microcanonical mean directly.

One of the most influential aspects of Gibbs's book is what may be called his pragmatist attitude to statistical mechanics. Rather than "attempting to explain the mysteries of nature," Gibbs wrote, he was "forced to be contented with the more modest aim of deducing some of the more obvious propositions relating to the statistical branch of mechanics. Here, there can be no mistake in regard to the agreement with the facts of nature."[31] Gibbs emphasized that statistical mechanics, like other physical theories, was an intellectual construction and not a simple mirror of reality.[32] That is, the value of statistical mechanics should be measured solely, or at least mainly, by its ability to agree with our empirical laws of thermodynamics and by its ability to explain them. For these purposes it is not necessary that each term in the theory

[30] For Gibbs's treatment of the microcanonical distribution, see his *Elementary Principles*, ch. X.

[31] See ibid, p. x.

[32] A context where Gibbs's general pragmatist temperament is clearly visible is in his discussion on the so-called "Gibbs Paradox." When we mix two portions of gas that are identical in all of their thermodynamical parameters, we may calculate the entropy by using the generic phases (in which case the entropy will not increase after the mixture) or the specific phases (in which case the entropy will increase.) Many physicists regard the fact that these two calculations are not identical as a paradoxical result. For Gibbs, however, the situation demonstrated how, in the course of doing physics, one needs to choose between intellectual constructions, each of which is perfectly acceptable in its own right. For a discussion on Gibbs's paradox see Brush, *Statistical Physics*, p. 75. Note that a solution of the paradox may be obtained if we regard the gas particles as indistinguishable. This solution, though, is acceptable only in the context of quantum mechanics, where the indistinguishability of the particles is assumed on independent grounds.

have a matching "element of reality".[33] In particular, Gibbs rejected the view that statistical mechanics is based on an atomic hypothesis. "Certainly," wrote Gibbs, "one is building on an insecure foundation, who rests his work on hypotheses concerning the constitution of matter."[34] Because of his pragmatism, it was easy for Gibbs to accept probabilities as parts of the basic makeup of his physical theory. He assumed that statistical mechanics has "built-in" probabilities, so to speak, and expected his assumption to be considered justified if the resulting theory proved useful. This conception of probabilities should be distinguished from others. On the one hand, Gibbs's probabilities are not grounded in an assumption of a molecular disorder. Gibbs dispensed with Boltzmann's efforts to introduce probabilities through an appropriate mechanical stosszahlansatz assumption. (In a sense, Gibbs continued the line of reasoning that Boltzmann introduced in his combinatorial proof.) On the other hand, Gibbs did not conceive of probabilities as being external to mechanics. On the contrary, more than anybody else he is responsible for the realization of Maxwell's intentions, that is, to introduce probabilities directly into physics rather than regarding them as heuristic devices that we employ because of our lack of precise knowledge.

Note again that Gibbs's pragmatic conception of SM relied, crucially, on his use of the theory of errors. At each point he demonstrated that his oversimplified predictive model was justified because the various deviations from the model were distributed as if they were further sources of error. These demonstrations together prove that the accuracy of the model does not depend on the minute details. As a result, we may employ fairly crude models and reach very accurate results. Moreover, Gibbs's argument provides a reason against the use of a more refined and detailed model. (Indeed, it seems rather certain that even if we had a precise knowledge concerning the state of all of the molecules, we would still employ Gibbs' ensemble method.) In science as well as in painting, the inclusion of unnecessary details often has an obscuring rather than an illuminating effect.

As we shall see later in this book, a more rigorous demonstration of Gibbs's argument necessitates the application of the central limit

---

[33] This term echoes Einstein's terminology, but it may be given a broader interpretation.
[34] Gibbs, *Elementary Principles*, p. x.

theorem. In this connection, one should regard Khinchin's as a further continuation of Gibbs's project. Using the central limit theorem, Khinchin demonstrated that the details of the initial distribution of the energy between the particles need not be taken into account. In the thermodynamic limit, that is, when we are dealing with systems with infinitely many particles, the details become simply irrelevant. We shall return to this issue in the next and in the last chapters.

Gibbs's methods for calculating probabilities proved enormously successful. In a wide variety of contexts, calculations based on Gibbs's canonical and grand canonical averages agree completely with the observed frequencies of various events. Moreover, the problem of specific heats, a problem that Gibbs regarded as the main defect of his theory, was solved shortly after Gibbs's death.[35] Gibbs's theory also aged well in other respects. One may even say that the Gibbsian formulation of quantum statistical mechanics is even more fruitful and indispensable than Gibbs's own classical statistical mechanics.[36] (In the context of quantum mechanics, other competing versions of SM seem a lot weaker.) These success stories provide ample reasons for regarding Gibbs's theory, complete with its probabilistic assumptions, as true. Indeed, many physicists have no qualms with this pragmatist justification of Gibbs's methods.[37] Nevertheless, as a matter of historical fact, Gibbs's theory did not pacify the doubts concerning the physical reality of the statistical ensembles any more than William James's pragmatic treatment of religion[38] succeeded in providing a universally

---

[35] See Kuhn, *Black Body Theory.*
[36] For a treatment of quantum SM, see, for example, R. C. Tolman's *Principles of Statistical Mechanics*, Claredon (Oxford, 1938) pp. 325–362.
[37] See ter Haar 's "Foundations of Statistical Mechanics," *Rev. Mod. Phys.* 27(3) (1955), for a discussion of the pragmatic justification of Gibbs's methods.
[38] For William James, as for Gibbs, the question about the existence of an entity should be settled when it is demonstrated that the belief in the existence of the disputed entity is useful. In particular, the belief is justified when it results in a theory with better explanatory powers. This applies to both the question about the existence of God as well as questions about the physical reality of the ensembles. James agreed with Leuba's remark that "God is not known, he is not understood; he is used. . . . If he proves himself useful, the religious consciousness asks for no more than that." See James, *The Varieties of Religious Experience*, Longmans, Green and Co. (New York, 1902) p. 502. By inserting "ensemble" instead of "God" and "scientific consciousness" insead of "religious consciousness," we obtain a good description of Gibbs's view of the reality of ensembles.

acceptable basis for religious beliefs.[39] This fact became especially clear after the near universal acceptance of the atomic conception of matter that followed the research of Einstein, von Smoluchowsky, and Perrin on the Brownian motion. The research on the Brownian motion proved, beyond the shadow of a doubt, that the introduction of probabilities into SM was indeed necessary because the second law of thermodynamics has only statistical validity. That realization, in turn, renewed the interest in questions concerning the physical reality of the ensembles.

There are signs that even Gibbs himself had some doubts concerning his pragmatist attitude. In the twelfth chapter of *Elementary Principles*, Gibbs remarked that his ensembles may be understood, ultimately, as the collection of states into which a single system evolves in an infinite duration of time.[40] This conception was much more congenial to Einstein, for example, because the totality of future and past states of a single system is (potentially) an observable totality.[41] The totality of identical but differently prepared systems (the usual definition of an ensemble) is not an observable totality at all (that is, unless the mode of preparation is specified in detail). Hence, one gains a more robust physical content when attention is restricted to the evolution of a single system. Notice that if the notion of an ensemble can be replaced with the notion of the future and past states of a single system, then the concept of the probability of an event A could be replaced by the relative amount of time that a single system spends in states with the property A. This idea gave rise to the ergodic approach that is the better known alternative to the pragmatist conception of SM. This approach will be the center of the discussion in the next chapter.

In this chapter we shall be preoccupied with another criticism of Gibbs. This is an attempt to move the clock back and return to a Laplacian conception of probabilities. As you may recall, Laplace

[39] For some, the main problem was that probabilities were not defined by Gibbs as observable quantities, while for others the question was whether statistical mechanics was consistent with the more fundamental principles of physics. We shall return to these issues later in the book.

[40] This idea appeared even earlier in the writings of Maxwell, who took it from Boltzmann. (See Boltzmann, "Weitere Stidien Uber das Warmegleichgewicht Unter Gas Molekulen" from 1872.) For a discussion of Gibbs's notion of ensembles, see von Plato's *Creating Modern Probability*, p. 82.

[41] For an excellent discussion on Einstein's views on probability, see von Plato's, *Creating Modern Probability*, pp. 114–123.

believed that we employed probabilities solely because of our lack of certain and precise knowledge. Probabilities allow us to make predictions even when our knowledge is partial and only approximately correct. According to the Laplacian conception, then, probabilities in general, and the law of the distribution of errors in particular, are not part of physics proper. Physics attempts to give an objective description of the world. Probabilities, on the other hand, are merely inferential devices. According to Laplace, an omniscient being will have a complete picture of the world without mentioning probabilities at all. For Laplacians, then, an honest approach to SM should emphasize its inferential role and present the probabilities as degrees of certainty.

Laplace's view on probabilities was motivated by his determinism. The view that the state of the world at one instance determines its state at any other moment does not leave any room for randomness, stochastic processes, or any other objective statistical concept.[42] But even those who have a skeptical view toward determinism may, nevertheless, agree with the thrust of Laplace's views on probabilities. Many modern thinkers begin with Laplace's conclusion, that is, they maintain that probabilities are subjective attitudes formed because of ignorance and are devoid of any objective meaning. These thinkers do not necessarily subscribe to any form of determinism, but they nevertheless attempt to eliminate objective probabilities and replace them with degrees of belief, degrees of confidence, or simply with betting rates. By using such concepts, one may attempt to reconstruct SM in subjectivist terms.

The rest of this chapter will center around the work of E. T. Jaynes, who developed a subjectivist view of SM in his two-part article entitled "Information Theory and Statistical Mechanics"[43] (henceforth SMIT I,II). Jaynes attempted to show that the predictive capacity of SM is largely independent of the deterministic character of its dynamics.[44] Jaynes's conception of SM is rather unique; indeed, he refers to the subject of his work as "predictive statistical mechanics," to

---

[42] It did, however, leave room for a discussion on instability, which he conducted in his treatise on celestial mechanics.

[43] See E. T. Jaynes in *Papers on Probability, Statistics and Statistical Physics*, ed. R. D. Rosenkrantz, Reidel (Dordrecht and Boston, MA, 1983), pp. 4–39.

[44] This point is important for several reasons. First, this way of thinking makes it easier to see how classical and quantum SM can be unified. Second, the view that probabilistic reasoning should be divorced from questions concerning determinism was taken by de Finetti and other subjectivists as an important addition to their

distinguish his line of thinking from others'. The differences run rather deep. To begin with, Jaynes says that "there is no necessary connection (between predictive SM) and thermodynamics, other than the fact that thermodynamics provided, historically, the first example of such a problem."[45] Moreover, when he cites examples of the use of predictive SM he mentions, among other examples, image reconstruction as belonging to this discipline. Another point that makes Jaynes's approach unique is that he does not regard predictive SM and statistical inference as different subjects. "The former," he says, "is only a particular realization of the latter. It applies equally well in any generalized inverse problem in which we have prior information on multiplicities that need to be taken into account."[46] Whereas many physicists will readily admit that they use statistical methods with applications outside of physics, most will hesitate to characterize the whole discipline of SM in such general terms. Indeed, the extreme nature of Jaynes's attempt to divorce SM from its particular scientific background motivates us to call his view ultrasubjectivist, to distinguish it from other, less radical attempts to provide epistemic foundations for SM probabilities.

Jaynes's thesis has a fairly long history.[47] Maxwell started a tradition when he introduced an observer with limited discriminatory capacities into SM with the concept of a coarse-grained partition of the state space or phase space of a mechanical system. In addition, Gibbs's and Boltzmann's concept of an ensemble of differently prepared systems of the same type over which a host of probability functions could be defined also contains many of the ingredients of subjectivism with regard to SM.[48] Other more recent contributors to this tradition include Fowler,[49] Tolman,[50] and ter Haar.[51] However, it was Jaynes who insisted that the various disagreements over the foundations of SM belonged

---

epistemic interpretation of probability. In this respect, Jaynes's view is more faithful to the subjectivist spirit than the subjectivist version of the ergodic approach, which we will present in the next chapter.

[45] See E. T. Jaynes, ibid, p. 2.
[46] Ibid.
[47] For Jaynes's own version of the history of his views, see his "Where Do We Stand on the Maximum Entropy Principle?", ibid, p. 210.
[48] According to Jaynes, his original intention was to provide an interpretation of Gibbs's laws. Ibid, p. 3.
[49] See, for example, Fowler's *Statistical Mechanics* (Cambridge, 1936).
[50] See Tolman, *The Principles of Statistical Mechanics*, ibid.
[51] See ter Haar's *Statistical Mechanics* , Reinhardt and Co. (New York, 1954).

to a more general genus – they are the consequences of the general disputes over the foundations of probability – and in this discovery lies the value of Jaynes's work.

To be sure, Tolman realized long before Jaynes that the rivalry between the ergodic approach and his own had a broad methodological background.[52] Ter Haar was even firmer on this point when he insisted that SM should be regarded as an instrument for correct predictions.[53] Jaynes, however, was the first to regard the a priori probabilities that Tolman introduced into SM as subjective probabilities, and he was also the first to call the general prediction process that ter Haar described a statistical inference. By doing so, Jaynes introduced the study of the foundations of probability to the community of physicists, and, at the same time, he made philosophers and statisticians realize the consequences that their views might have on SM.

Focusing our inquiry primarily around Jaynes, though, is not unproblematic. Shortly after he published SMIT, Jaynes began to refer to his view as an objectivist one.[54] Presumably, he did this to fend off the panic that the term subjectivism created among physicists.[55] But even if his view remained essentially unchanged, there is enough in a name to create genuine confusion. More generally, Jaynes's later work contains disclaimers of some of his earlier views with no explanation for his change of mind. This presents the reader with an unpleasant dilemma, because one must decide whether to address views that Jaynes himself presumably no longer holds or whether to do Jaynes's explanatory work for him. Though neither possibility is very appealing, we have chosen the first option, because we believe that Jaynes's original articles may be addressed on their own merits. Jaynes's original thesis

---

[52] See, for example, Tolman, ibid, p. 54. Von Neumann, too, was aware of the fact that the discussions on the character of SM probabilities had a broader philosophical background. See von Heumann, *The Mathematical Foundations of Quantum Mechanics*, Springer-Verlag (Berlin, 1932), pp. 2 and 108, for a discussion on the foundations of probability. See also the discussion in Gibbs, *Creating Modern Probability*, p. 112.

[53] See ter Haar, "Foundations of Statistical Mechanics," *Rev. Mod. Phys.* 27(3) (1955).

[54] In his 1983 introduction to SMIT I, Jaynes writes: "I no longer subscribe to the views about 'subjective' and 'objective' probability theory on p. 8." See *Papers . . .* , p. 4. As early as his 1963 "Brandeis Lectures" (ibid, pp. 40–76), Jaynes put parentheses around the terms objective and subjective. Later he called his view an "impersonalist" view. I have not been able to determine whether it was Jaynes himself or his commentators who coined the term "objective Bayesian" to describe Jaynes's view.

[55] Ibid, p. 18.

can be seen as either the provocative conclusion of an interesting tradition within SM or as an application of provocative ideas from statistics to SM. Viewed in either way, Jaynes's ideas are interesting and important even if we find that, ultimately, they deserve serious criticism.

Throughout the rest of this chapter we shall refer mainly to SMIT I,II. Consequently, we shall not mention many of the themes that became central to Jaynes's theory after SMIT. In particular, we shall not address a controversial issue that seems to be very important to Jaynes's current concerns, namely, the claim that the maximum entropy principle he advocates (henceforth MEP) is as defensible as any logical principle. It seems that the criticisms of Shimony[56] and Seidenfeld[57] shed serious doubt on this thesis. However, insofar as the maximum entropy inference can be presented as an inference that leads to correct results in the context of SM, it seems to us to deserve serious attention. More generally, Jaynes's view articulates an important undercurrent that exists in the writings on the foundations of SM.

1.5. THE PREDICTIVE APPROACH TO SM

As we mentioned earlier, we believe that Jaynes's theory should be presented as a response to Gibbs's pragmatist approach. However, the immediate source of the ideas to which Jaynes reacted was not Gibbs himself but Tolman, who canonized Gibbs's approach in a book he wrote in 1938. In many respects, Tolman's book was not a mere presentation of Gibbs's ideas. Tolman's presentation is more rigorous mathematically and more careful methodologically; it also includes an application of Gibbs's ideas to quantum mechanics. In addition, Tolman was completely confident in the empirical adequacy of statistical mechanics whereas Gibbs, who was worried about the issue of specific heats, was much more hesitant to declare statistical mechanics victorious. Tolman introduced the ensemble method almost at the outset. He did so in the following way:

[56] See Shimony and Friedman, "Jaynes Maximum Entropy Prescription and Probability Theory," *J. Statist. Physi.* 3(4) (1971). Shimony's "Comment on the Interpretation of Inductive Probabilities," *J. Statist. Phys.* 9(2), and "On the Status of the Principle of Maximum Entropy" *Synthese* (1983). See also Hobson, "The Interpretation of Inductive Probabilities," *J. Statist. Phys.* 6(213) (1972).

[57] T. Seidenfeld, "Why am I not an Objective Bayesian," *Theory and Decision* 11 (1979) and "Entropy and Uncertainty," *Phil. Sci.* 53 (1986).

We are now ready to begin our consideration of the statistical methods that can be employed when we need to treat the behavior of a system concerning which condition we have some knowledge but not enough for a complete specification of the precise state. For this purpose we shall wish to consider the average behavior of a collection of systems of the same structure as the one of actual interest but distributed over a range of possible states. Using the original terminology of Gibbs we may speak of such a collection as an ensemble of systems.[58]

Each member of the ensemble is represented as a phase point. (That is, if the system of interest has n particles, the members of its ensemble will be represented as points in a $6n$-dimensional space. Such a representation encodes for each particle three coordinates of position and its three coordinates of momentum.) Further explaining the notion of an ensemble, Tolman writes:

In using ensembles for statistical purposes, however, it is to be noted that there is no need to maintain distinctions between individual systems since we shall be interested merely in the number of the systems at any time which would be found in the different states that correspond to different regions in the phase space. Moreover, it is also to be noted for statistical purposes that we shall wish to use ensembles containing a large enough population of separate members so that the number of systems in such different states can be regarded as changing continuously as we pass from the states lying in one region of the phase space to those in another. Hence, for the purposes in view, it is evident that the condition of an ensemble at any time can be regarded as appropriately specified by the density r with which representative points are distributed over the phase space.[59]

When we normalize r we obtain a probability measure over the phase space. This probability function is to be interpreted as assigning a "relative weight" to those members of the ensemble that lay in a given infinitesimal region of the phase space at a given time. As time progresses, each member of the ensemble follows its deterministic route and traces a trajectory in the phase space. The Gibbsian point of view, though, concentrates only on the way the density r changes in time, that is, on the quantity $\partial r/\partial t$. This quantity summarizes

[58] See Tolman, *Principles of Statistical Mechanics*, Clarendon (Oxford, UK, 1938), p. 41.
[59] Ibid, p. 46.

all of the relevant dynamic factors that we need to describe how the ensemble averages change in time. Concerning the choice of r, Tolman writes:

Some kind of postulate as to a priori probabilities is always involved in the use of statistical methods for predictive purposes. . . . For the above purposes we now introduce the hypothesis of equal a priori probabilities for different regions in the phase space that correspond to extensions of some magnitude. . . . As already emphasized, this principle should be regarded as a postulate. Nevertheless, we can see the reasonable character of this principle if we consider the behavior of the uniform ensemble. . . . If we set up an ensemble with the phase points for its member systems uniformly distributed, with the same number in all equal regions of the phase space without preference to location, this uniform density of distribution would be permanently maintained as time proceeds. We thus find that the principles of mechanics do not themselves include any tendency for phase points to concentrate in particular regions of the phase space.[60]

Ultimately, however, Tolman does not regard the invariance of the density in time, a consequence of Liouville's theorem, as the only justification of the postulate of equal a priori probabilities. When summarizing his point of view, Tolman talks about the simplicity and the empirical success of the hypothesis as the grounds for accepting it:

The methods (of statistical mechanics) being statistical in character have to be based on some postulate as to a priori probabilities and the hypothesis chosen is the only postulate that can be introduced without proceeding in an arbitrary manner. The methods used lead to results which do agree with the empirical finding.[61]

Concerning the relations between Gibbs's view and his own, Tolman wrote:

The above point of view does not conflict with that of Gibbs. He was much less explicit, however, concerning the hypothesis of equal a priori probabilities for equal regions in the phase space and much less confident about the validity of statistical mechanics, presumably because of its failure to account for phenomena – such as specific heats – which we now know must be treated by the methods of quantum rather than classical statistical mechanics.[62]

---

[60] See Tolman, p. 59.   [61] Ibid, p. 65.   [62] Ibid, p. 65.

Tolman was also very explicit in his rejection of the idea that, ultimately, the ensemble averages should be presented as a convenient substitute for the time averages of the system of interest. When he assessed this idea, which is the basis of the ergodic approach, he wrote that such a point of view "will prevent the methods of statistical mechanics from exhibiting their full statistical character and usefulness."[63] Tolman also pointed to the problematic nature of the idea that we may identify time averages taken over an infinite time with the "long-term behavior" of the system of interest.

Tolman's point of view remained influential for a long time. Nevertheless, the idea of justifying statistical mechanics in a pragmatic manner did not remain unchallenged. First, some physicists, notably Uhlenbeck, remained convinced by the ergodic approach. But there were other dissenting voices as well. Take, for example, the following passage from an important review article by ter Haar that Jaynes mentions in the beginning of SMIT:

One might argue that the proof of the pudding is in the eating and that the fact that statistical mechanics has been able to predict accurately and successfully the behavior of physical systems under equilibrium conditions – and even under certain non-equilibrium conditions – should be sufficient justification for the methods used. This is especially the case nowadays since the introduction of quantum mechanics has cleared up such seemingly insurmountable difficulties as the paradox of specific heats or Gibbs's paradox.[64]

However, it is only logical that many physicists have not felt satisfied by this argument and have therefore attempted to show that the statistical formalism follows in a straightforward way from classical or quantum mechanics."[65]

As an alternative to the pragmatist approach, ter Haar mentions the *H*-theorem approach, which he characterizes as the approach that emphasizes predictive issues, and the ergodic approach. Jancel, in another important book on statistical mechanics (*Foundations of Classical and Quantum Mechanics*, Pergamon Press, New York), also distinguishes between the utilitarian, the formal, and the physical points of view. (These are somewhat idiosyncratic terms for, respec-

---

[63] Ibid, p. 68.
[64] See footnote 31 for a remark on Gibbs's paradox.
[65] See ter Haar, "Foundations of Statistical Mechanics," p. 289.

tively, the pragmatist, the ergodic, and the predictive approaches.) But the distinction between the pragmatic and the predictive approaches remained somewhat unclear before Jaynes. Indeed, because both Gibbs and Tolman emphasized the fact that the task of statistical mechanics is to make predictions, it is easy to be confused about the distinction to which we are alluding. Let us, then, see how Jaynes introduced his approach to statistical mechanics that we termed the neo-Laplacian approach.

The simplest and perhaps the most important point that Jaynes made is that, if one is interested in questions of predictions or inferences in the context of statistical mechanics, one should discuss these statistical inferences within a more general framework, because only in such a framework can we characterize the inferential process itself, rather than the specific selection of inferential problems we happen to chance upon. With this aim in mind, one should look at statistical mechanics from the broader perspective of the mathematical theory of probability. Such a perspective may enable us to give a more precise and abstract characterization of the inferential problems we encounter in physics. This point may sound almost trivial, and therefore the extent to which physicists have resisted it is rather astounding.[66]

The second preliminary point that Jaynes made is a direct continuation of the first. When one attempts to characterize statistical inference, one should be attentive to the broader methodological issues connected to the definition of probability. Again, the discussion on the nature of physical probabilities had remarkably little references to the more general literature on the foundations of statistics. Jaynes emphasized the fact that some of the problems that physicists have been grappling with have very little to do with the specific background of statistical mechanics. Therefore, even in the context of physics, the full range of possible approaches to the theory of probability should be explored.

Among those who wrote on the foundations of probability, Jaynes mentions three writers who influenced his views. The first is Laplace, who insisted that probabilities should be viewed as degrees of belief. The second writer often mentioned by Jaynes is Harold Jeffreys, who

---

[66] To take one example among many, it is rather puzzling to see how Darwin and Fowler reinvented for themselves whole portions of the theory of probability. Khinchin, who used the limit theorems of modern probability theory, managed to arrive at precise and more concise formulations.

showed how to formulate the constraints we impose on our degrees of belief in a general mathematical way. (In many cases, such constraints can be formulated as the requirement that our degrees of belief should be invariant under the action of a specific transformation group.) The third writer is Shannon, the inventor of information theory. Shannon showed how to extend Laplace's principle of insufficient reason and apply it to new contexts. In addition, Shannon's methods may be used to justify the principle of insufficient reason in a new way. The above-mentioned principle embodies a commitment to the principle that one ought to be maximally nonjudgmental with respect to missing information.

Using the insights of these writers, Jaynes attempted to characterize the inferences that physicists implicitly make when they use Gibbs's rules. The picture he arrived at was the following. When we make predictions concerning the state of a mechanical system, we have to go through the following stages. First, we have to articulate all of the information we have concerning the system. Second, we have to enumerate all of the hypotheses concerning the precise state of the system that are compatible with what we know. Third, we have to assign subjective probabilities to each of these hypotheses. The probabilistic assignments are often constrained by some general principles. Finally, when we try to calculate the probability that the system has some property, we have to assess the "relative weight" of those states that are compatible with this property. As we shall see shortly, these stages can be viewed as another way of describing Tolman's account of what is involved in constructing an ensemble – fixing the density measure and calculating the expectation values. Nevertheless, there is at least one serious difference between Tolman's approach and Jaynes's. Jaynes regards the process whereby we summarize the information concerning the dynamics of the system as a species of a more general genus, namely, the deliberate effort to simplify prediction problems by ignoring information deemed irrelevant. For Tolman and Gibbs, the choice of a "thinner" description of many-particle systems is one of the aspects of statistical mechanics that may be regarded as *justified* when the theory as a whole is proven fruitful and empirically correct.

In summary, let us say that there are two approaches that aspire to replace Gibbs's pragmatist approach to SM – the ergodic approach and the predictive approach. The former is a traditional approach that attempts to give dynamic explanations to the puzzles of SM, and the latter is a revisionist approach which provides explanations that depend only on very general dynamic considerations. The main tenet of the predictive approach (and one that is consistent with Gibbs's approach) is that the large number of degrees of freedom is responsible for equilibrium phenomena. It is an unorthodox idea that similar systems that differ only in the number of their degrees of freedom have to be distinguished by physics, because the change in number does not seem to bring any new physical considerations. Because of this unorthodoxy, many writers on the subject felt the need to present the predictive approach in a broader methodological light. Jaynes chose subjectivism as his general methodology, a choice that seems appropriate because for subjectivists, who are generally interested only in the question of prediction, the shift from dynamic considerations to questions concerning prediction is welcome. Therefore, when the probabilities are interpreted as subjective conditional probabilities, the shift from the objective dynamic questions concerning the behavior of the system "in and of itself" to issues concerning what an agent may predict about the system can acquire a clearer motivation. This shift can be seen as an instance of the general subjectivist tendency to recast scientific concepts in epistemic terms. Dynamic considerations may seem superfluous when they are not related to those experimental facts we are interested in predicting. So the reason for the subjectivists' resistance to such considerations is clear. And, once the subjectivist framework is introduced, the predictive approach seems more natural because the large number of degrees of freedom certainly affects our ability to make predictions.

This last point may serve to explain why we refer to Jaynes, and those who share his views, as ultrasubjectivists. In the first place, Jaynes's theory plays down the importance of the specific background of statistical mechanics. Second, one of the outcomes of Jaynes's treatment is a demonstration that, for the purposes of prediction, the deterministic assumptions do not play an essential role. This type of elimination program was advocated explicitly by de Finetti,[67] who believed that questions about determinism should be divorced from those concerning foresight (his term for the prediction of observable events). And because he was a subjectivist, he claimed that he was only interested in foresight. In fact, de Finetti thought that this line of thinking was in the tradition of Hume. De Finetti, like Hume, emphasized

[67] See, for example, de Finetti's "Probabilism," *Erkenntnis* 31 (1989), pp. 169–223. (Original publication 1931.)

the connections the mind makes between ideas [for de Finetti the strength of the connection between A and B is expressed by the conditional probability $P(A/B)$] and both doubted the objectivity of causal laws.[68]

In the next sections we shall present systematically the ideas of the predictive approach. We shall show how to present it as an eliminativist program and criticize its shortcomings.

1.6. THE CONCEPT OF EQUILIBRIUM

Of the different problems associated with the foundations of SM, the ones that attract the most attention are those connected with the definition and explanation of thermodynamic equilibrium. The first problem is that thermodynamic processes such as solidification, evaporation, or the creation of magnets and crystals are definable only as macroscopic phenomena. However, we expect there to be a way of accounting for thermodynamic phenomena on the microscopic level as well. This, though, is not an easy task. From the macroscopic point of view, these types of processes are extremely stable and regular, and this stability is hard to reconcile with the fact that we conceive of SM systems as being composed of seemingly irregularly moving molecules. As we mentioned earlier, the key to the solution of this problem is the idea that the macroscopic descriptions refer to the "most probable states" from a microscopic point of view. Hence, all of the problems connected with the concept of probability become issues that are directly relevant to SM.

The second problem that one faces in attempting to define and explain thermodynamical equilibrium is that thermodynamic systems are not only stable, but also irreversible. We do not expect, for example, that, upon being stirred, the milk will spontaneously separate from the coffee. The possibility of that happening, however, is compatible with the microscopic description of the milk and coffee system. Furthermore, we have a theorem – Poincaré's recurrence theorem – that states that in the infinitely long run the system is expected to return arbitrarily close to its original state. Here too we solve the problem by

---

[68] De Finetti wrote: "Certainly, we cannot accept determinism; we cannot accept the 'existence' in that alleged realm of darkness and mystery, of immutable and necessary 'laws' which rule the universe, and we cannot accept it as true simply because, in the light of our logic, it lacks all meaning." Ibid, p. 169.

introducing probabilities; we tend to say that spontaneous separation is possible but very rare in a probabilistic sense, which again amounts to delegating the solution of the problem to the probabilist. Note that under certain conditions reversals do actually happen.[69] Then, a third problem that one faces is how to explain thermodynamics without explaining away improbable but nevertheless real phenomena. These three problems must be solved by those interested in the foundations of SM; the different proposed solutions to these problems are intimately connected to different views on the foundations of probabilities in general. Objectivists try to introduce external and real parameters to account for the thermodynamic phenomena, whereas the subjectivists maintain that no such probabilities need to be introduced. In this second category we find both permissive subjectivists, who attempt to reconcile subjectivism with the necessary dynamic assumptions, and ultrasubjectivists, who maintain, as Jaynes put it, that "statistical mechanics need not be regarded as a physical theory whose validity depends on the truth of additional assumptions not contained in the laws of mechanics."[70] In other words, they will argue that the probabilistic arguments that we find in SM are justified on a priori grounds.

To fix ideas, let us return to the ideal gas case, but this time we shall attempt to be more precise. Let M be a system of n particles that are contained in an isolated container. We assume that the intermolecular forces are negligible. A mathematical description of a state of M can be given as an n-tuple $r_1, \ldots, r_n$ of points in the six-dimensional Euclidean E^6 space such that r_i represents the three positions and the three momentum coordinates of the ith particle. This description is different from the phase space description, where a state of M is identified with a single point in the $6n$-dimensional space. There is, however, a clear correspondence between the two types of descriptions.[71] The fact that M is located in an isolated container can be translated into the

---

[69] The spin echo effect proves that there are cases where systems might "remember" their previous ordered state and regain it after assuming a disordered state. For a popular article on the spin echo effect, see R. G. Brewer and E. I. Hahn, "Atomic Memory," *Sc. Am.* (1985), pp. 42–49. For a discussion on the philosophical significance of the effect, see Sklar, *Chance and Physics*, Cambridge Univ. Press (New York, 1993), pp. 219–222. For an article by Jaynes on related issues, see "The Minimum Entropy Production Principle," pp. 402–424.

[70] Jaynes, "Papers . . . ," p. 6.

[71] To each ordering of $r_1, \ldots, r_n$ there corresponds a point in the $6n$-dimensional space. For more details, see Uhlenbeck and Ford, *Lectures on Statistical Mechanics*, American Mathematical Society (Providence, RI, 1963), pp. 1–7.

assumption that all of the states possible for M are located in S, which is a bounded connected subspace of E^6. (The boundedness in position space is a result of the walls of the container. The boundedness in the momentum space is the result of the fact that the system is in a given total energy and is isolated.) Let $P = P_1, \ldots, P_m$ be a finite partition of S. We shall think of the P as our "descriptive framework" for M. The number m is assumed to be much smaller than n, but it should be large enough so that if we know in which member P_i a particle r_j is located, such knowledge will be a "good approximation" of the exact state of r_j. Therefore, given a state $r_1, \ldots, r_n$ we may partially describe the state by the numbers $n_1, \ldots, n_m$ such that n_i, $0 \le i \le m$, is the number of particles located in P_i. n_i is known as the occupation number of P_i.

Now suppose that, initially, all of the particles of M were located in a small compartment of the container. This means that $n_i \ne 0$ only for a small number of values of i. Then the gates of the compartment are opened and the particles float freely in the container. In terms of the present example, the equilibrium will be reached when the n particles are dispersed throughout the container. The foundational problem of SM is to explain why such equilibrium will always be obtained after a while. More precisely, the following questions arise: (1) Why do we expect the molecules to disperse throughout the container after a while? (2) Why do we expect the distribution of the momentum among the particles to be normal? (3) Why do we expect this "equilibrium" to be stable? That is, why do we expect the particles to remain dispersed in position and momentum? According to the terminology we have developed, our expectations can be rephrased in the following ways: (a) We expect $n_j \ne 0$ for all j. (b) We expect the ratio between n_i and n_j to be proportional to the volumes of the areas of P_i and P_j. (c) If, for all j, the areas of P_j are the same, we expect little variation between n_j and n/m. In addition, we have to explain why we expect this situation to remain stable as time goes by. As we remarked earlier, Boltzmann first attempted to answer this question by articulating the idea that the dynamics of most of the n-particle systems will lead them to evolve to a state that answers the above expectations. Let us see, however, how Boltzmann approached the situation in his 1868 paper.[72] The key to Boltzmann's idea is that the macroscopic description of an equilibrium

[72] Our presentation in this section will follow Uhlenbeck and Ford, ibid, rather closely.

state does not distinguish between many of the microscopic states that are compatible with it. Moreover, he showed that if the equilibrium is obtained when the particles disperse throughout the container, then the overwhelming majority of the microscopic states are compatible with the set of equilibrium states. More precisely, because a set of occupation numbers $n_1 \ldots n_m$ does not distinguish between many of the exact states of the n-particle system, we can ask the following question: What is the phase volume $W(n_1 \ldots n_m)$ of the set of points that represent states between which the occupation numbers $n_1 \ldots n_m$ do not distinguish? (1) Because a motion of a particle inside the box P_j does not change the occupation numbers, and (2) because a permutation of two particles does not change the occupation numbers of their respective cells, it is easy to see that the volume in the phase space of the exact states that correspond to one and the same set of occupation numbers is proportional to:

$$ W(n_1 \ldots n_m) = \frac{n!}{n_1! n_1! \ldots n_m!} u_1^n \ldots u_m^n, $$

where u_i is the volume of P_i. [The magnitude $n!/(n_1! n_2! \ldots n_m!)$ is the number of states obtained from a given state by a permutation of two particles, and $u_1^n \ldots u_m^n$ is, roughly, a measure of the number of states obtained from a given state by moving a particle inside the box P_j.]

Now we can ask a further question: What are the conditions on $n_1 \ldots n_m$ that will determine whether $W(n_1 \ldots n_m)$ is large or small? Or, better still, are there any conditions on $n_1 \ldots n_m$ that will make $W(n_1 \ldots n_m)$ overwhelmingly large (relative to the total volume in phase space of the states possible for M)? The somewhat surprising answer to the last question is that if the volumes $u_1 \ldots u_n$ of $P_1, \ldots, P_m$ are equal and n is very large, the occupation numbers $n_1 \ldots n_m$ that answer our expectations (a), (b), and (c) approximately yield an overwhelmingly large area $W(n_1 \ldots n_m)$. This is already beginning to sound like an explanation of our expectations; we expect the "most probable" case to happen! In other words, it seems that we have proof that the thermodynamic equilibrium is the "most probable" case. Observe that this argument is free of any dynamic assumptions.[73] It explains why, given that we do not know much about M, we should expect M to be in equilibrium. In particular, there is no use being made of the fact that in clas-

[73] See Jaynes, p. 227.

sical SM we are dealing with a deterministic system. In the next section we shall make this argument more precise and see how Jaynes used it for his ultrasubjectivist program.

1.7. THE PRINCIPLE OF MAXIMUM ENTROPY

Jaynes writes: "The only place where subjective statistical mechanics makes contact with the laws of physics is in the enumeration of the different possible, mutually exclusive states in which the system might be."[74] Jaynes is right to emphasize that enumerating the states possible for M presupposes a contact with the laws of physics: We are looking for those states that are possible *for M*, in an objective sense of possibility.[75] In fact, Jaynes tacitly makes another nontrivial assumption about the connection between the physical background and the subjectivized version of SM, namely, that the physical background is sufficient to determine the choice of the partition $P_1, \ldots, P_m$ uniquely.[76] As we shall see later, when the choice of partition is not derived from the background theory, the resulting inferences may be sensitive to the choice of partition. These, however, are the only objectivist assumptions that Jaynes is willing to make. In particular, if many writers tend to assume some correlation between the probabilities used in SM and the corresponding volumes in phase space, no such objectivist assumptions will be permitted by the ultrasubjectivist. Therefore, to derive the "overwhelming likelihood" of the equilibrium as we did in the last section, we need to account for the choice of $u_1 \ldots u_m$ as our subjective probability that the system will assume states compatible with $P_1, \ldots, P_m$, respectively, without introducing the concept of volume into our considerations. In this section we shall see how Jaynes proposed to achieve this goal. We shall introduce his argument that a principle for choosing $u_1 \ldots u_m$, which he calls the maximum entropy principle, can motivate the choice of the "right" $u_1 \ldots u_m$ without objectivist assumptions. Jaynes argued that his conclusion rested solely on "first principles," and claimed that there is a universally binding prin-

[74] See ibid, p. 6.
[75] The notion of "possibility for" is introduced and discussed by I. Levi, *The Enterprise of Knowledge*, MIT Press (Cambridge, MA, 1980). See, for example, Chapter 11.
[76] See Jaynes, p. 223.

ciple that motivates this choice. This principle is called MEP, the maximum entropy principle.[77]

Suppose that S is a set of states possible for M and $P_1, \ldots, P_m$ is a partition of S. Let $u_1 \ldots u_m$ be a possible assignment of probabilities to $P_1, \ldots, P_m$, that is, $\sum_{i=1}^{m} u_i = 1, 0 \leq u_i \leq 1$. If you have no information whatsoever concerning M, you may ask for a principle for selecting the $u_1 \ldots u_m$ that will somehow express the fact that you have no such information. In other words, you may ask whether there is a quantity $H(u_1 \ldots u_m)$ that obtains a maximum value when $u_1 \ldots u_n$ are the result of maximum uncertainty. Shannon suggested that if such a quantity existed, it would necessarily have the following features:[78]

1. H is a continuous function of the u_i.
2. If all of the u_i are equal, $H(u_1 \ldots u_m)$ taken as a function of m is monotonic increasing.
3. $H(u_1 \ldots u_m) = H(u_1 \ldots u_l) + H(u_{l+1} \ldots u_n)$.

(The motivation for the third feature is the idea that, by "regrouping" the different elements P_i, you do not change H.) Convince yourself (if you can) that these requirements are necessary. Somewhat surprisingly, these requirements determine H up to a constant! It can be shown that any function with the above properties can be written as $-K\sum_{i=1}^{n} u_i \log u_i$,[79] which is known as the entropy of $u_1 \ldots u_n$. Now, armed with the principle that one's subjective probabilities ought to be maximally entropic in general, Jaynes was able to derive the Maxwell–Boltzmann distribution that we constructed in the last section. That is, let $P_1, \ldots, P_m$ be the elements of a fixed partition of S. Let $u_1 \ldots u_m$ be your subjective probabilities that M will be found in a state compatible with $P_1, \ldots, P_m$, respectively. Assume that you are maximally ignorant with respect to M.[80] Assume also that you follow MEP and hence you

[77] Note that for subjectivists no principle above and beyond coherence is truly universally binding. The most you can say about other principles is that they are habitually used, intuitively appealing, etc. This is how de Finetti regarded exchangeability, for example. So, if Jaynes believes that the maximum entropy principle is "morally binding," as he sometimes puts it, he goes beyond the framework of ordinary subjectivism and that of ultrasubjectivism. We shall therefore address a variant of Jaynes's theory that does not take this extra step.

[78] Shannon's papers can be found in *The Mathematical Theory of Communication*, eds. C. E. Shannon and W. Weaver, Univ. of Illinois Press (Urbana, 1949).

[79] K is a constant.

[80] More precisely, suppose that there are no nontrivial features of M that we can reproduce in a laboratory.

choose u for which $H(u_1 \ldots u_m)$ is maximal; that is, you will choose u_1 $\ldots u_m$ for which the quantity $H(u_1 \ldots u_m) = -K\sum_{i=1}^{n} u_i \log u_i$ is maximal. It can be shown that the maximum of H is obtained when $u_1 \ldots u_m$ are all equal, and that is exactly what Jaynes wanted. The measure you choose for purely a priori reasons will be, in the end, very similar to the volume measure. (In technical terminology, it will be absolutely continuous with respect to the volume measure.) And that finally proves that our derivation of the overwhelming likelihood of the equilibrium state did not depend on an objectivist assumption about the underlying measure. According to Jaynes, this is the only obstacle that an ultrasubjectivist faces in the context of SM. Let us see how such an account would proceed.

Suppose that we observe an n-particle system M which has reached an equilibrium. To explain why it reached an equilibrium is, for Jaynes, to show why it was to have been expected or predicted. Such an explanation would be particular if particular information known about M were used for the prediction. If, however, no such particular information were used, the resulting explanation may be regarded as a general one. It provides a reason why systems of type M reach an equilibrium.[81] Let us see how Jaynes derived the predictions of an equilibrium state relative to any initial information. In most situations, the only information we have concerning the system M is derived from macroscopic measurements. Such information is so general, Jaynes maintains, that it does not rule out any state as a "seriously possible" future of M. Therefore, if $P_1, \ldots, P_m$ is a partition of the space S, Jaynes says that our probabilities $u_1 \ldots u_m$ (that M will assume a future state that lies inside $P_1, \ldots, P_m$, respectively) should be maximally noncommittal. For Jaynes this means that we ought to choose $u_1 \ldots u_m$ such that $H(u_1 \ldots u_m)$ will be maximal. But since $u_1 \ldots u_m$ corresponds in this case to the Maxwell–Boltzmann distribution, $W(u_1 \ldots u_m)$ (the phase volume of states compatible with equilibrium) is overwhelmingly large. That is, when n approaches infinity, the volume $W(u_1 \ldots u_m) = 1$, which means that (under the provisions alluded to above) we would predict an equilibrium state.

Jaynes derived the canonical distribution in a similar manner. As you may recall, the canonical ensemble is made of systems whose energy

---

[81] Note that to predict an equilibrium does not only mean to hold that it will happen with high probability. One actually forms the belief that the event will, in fact, happen. Hence, to put it starkly, we need to introduce into the discussion the reasoning we use to form a belief on the basis of our probabilities.

may be $E_1, E_2 \ldots$ or E_n but the mean of the energy $\mathbf{E}$ is fixed. Therefore, a probability distribution over the possible energy states $E_1, \ldots$ E_n has to have two properties:

1. $\sum_{i=1}^{n} p(E_i) = 1$

2. $\sum_{i=1}^{n} p(E_i)E_i = E$.

Gibbs found that one distribution that fulfills these conditions is

$$p(E_i) = e^{-bE_i} \bigg/ \sum_{i=1}^{n} e^{-bEj}.$$

This is the canonical distribution. He also noticed[82] that, of all the different distributions, this is the one that maximizes the expression

$$\sum_{i=1}^{n} p(E_i)\log p(E_i).$$

Jaynes argued that this point should be paraphrased in the following way. When the mean is specified, the canonical distribution is the only solution that is consistent with the PME. We shall return to this point later in this chapter.

Notice that, because we did not use any particular information on the system under consideration, this derivation is a general one. This point prompted Jaynes to regard the maximum entropy principle as a general principle of inference. According to Jaynes, the reasoning we apply in the context of statistical mechanics is the pure application of general principles of probabilistic reasoning. This is also the reason why we regard Jaynes's attitude as an ultrasubjectivist one. He tried to do away with any assumption that characterizes what is peculiar to SM.

After the publication of his 1957 article, Jaynes's views created a controversy among statisticians and physicists. For statisticians, the central issue was whether the maximum entropy principle is, indeed, an acceptable principle of inductive inference. Critics tried to show that the MEP suffers from the same problem as the principle of indifference.[83] Jaynes's counterargument was that, insofar as a principle of

[82] Gibbs, *Elementary Principles*, ch. XI.
[83] The best known of these problems is Bertrand's paradox. Put in a nutshell, the problem is that the probabilities do not remain the same when the problem is reformulated. For Jaynes's attempted solution to the problem, see his "A Well Posed Problem," ibid, pp. 133–150.

inference may have empirical support, the MEP is extremely well supported. For physicists, however, the issue was different. Jaynes's attack on the ergodic program was not only an attack on a very established view, but a demonstration that physics was not spared from philosophical difficulties. Specifically, he demonstrated that the opposition between the different schools of probability should be considered as an opposition within theoretical physics. It is still difficult for many to accept the fact that every controversy over issues concerning probabilities, be it the reference class problems, direct inference, or what have you, has a corollary within the context of SM.

1.8. THE H-THEOREM

In the last section, we learned how Jaynes argued that when we are practically ignorant about M we should expect it to be in an equilibrium state. Jaynes maintained that such a state of ignorance is not the exception but the rule, and as a result we usually find ourselves predicting that M will be in a state of equilibrium. Such predictions are routinely confirmed by experiments; however, there are some clear cases in which our predictions are not made relative to a state of complete ignorance about M. Therefore, the next question to raise is whether Jaynes's ultrasubjectivist analysis can be extended to cover such cases as well. In particular, there are two issues that are naturally raised: (1) When we know that M is in a state of equilibrium at t, why should we expect it to be in a state of equilibrium at t'? (2) When we know that M was prepared to be in a nonequilibrium state, why do we expect M to be in equilibrium after a while? To see how Jaynes proposed to answer these questions, we shall have to glance again into Boltzmann's and Gibbs's writings. But before doing so, let us mention another important point. The two above-mentioned questions seem to be, ostensibly, about inferential issues. However, if Jaynes could answer these questions successfully, he could claim that he had thereby also reformulated the issue of the "irreversible trend toward equilibrium" as an inferential rather than a dynamic issue. And the issue of irreversibility is what prompted Boltzmann to attempt and prove his H-theorem.[84]

---

[84] Boltzmann, as we mentioned earlier, made an assumption that the positions and velocities of the different particles of M are uncorrelated. On the basis of this assumption he set out to prove that, as a result of the effect of the collision between the particles, the entropy as a function of time will increase. The idea of his proof was that, for every scenario where a collision reduces the occupation number of a

The H-theorem is the name of an argument that Boltzmann developed in his 1877 paper. Boltzmann attempted to prove that once the Maxwell velocity distribution is attained, it will not be destroyed by further collisions. He did so by defining H, a function of the dynamic state of the system that coincides with Clausius' entropy (when the latter is defined), and proving that, as the system evolves, the value of H will always increase, attaining its maximum when the system arrives at the Maxwell distribution. The H-theorem, though, achieves more than what it was designed to achieve. First, it allows the notion of entropy to be generalized to nonequilibrium states. Second, the proof implies that even systems that are far away from equilibrium will become increasingly disordered. In this sense, the H-theorem is often taken as a general irreversibility proof that does not presuppose a trend toward an equilibrium. The H-theorem is generally regarded as the

---

given cell, we can construct a symmetric scenario where a collision will increase the occupation number. The proof proceeds by assigning equal probabilities to both scenarios. Boltzmann used an estimate of the expected average number of collisions that we can expect when the state of the system is not completely specified. Boltzmann's argument relied on the following assumptions:

1. The collisions with the walls of the container need not be taken into account. See C. Truesdell, *Fundamentals of Maxwell's Kinetic Theory of a Simple Monatomic Gas*, Academic Press (New York, 1980), pp. 152–160.
2. Only the collisions of two particles (rather than three or more simultaneous collisions) should be taken into account. (This is the so-called "diluted gas" case.)
3. The influence of external forces is negligible. (See ter Haar, ibid, on this point.)
4. The molecules of the gas are similar. (The so-called "simple gas" case.)

When one or more of these assumptions is removed, the proofs become much more complicated and the resulting arguments are more "realistic." What remains to be clarified, however, is the status of what has been proven because the assumption that the position and the momentum are not correlated was not given an independent justification. Much less is required for the purposes of prediction, because we can expect equilibrium to obtain in a very large range of cases, some of which cannot be dealt with in the Boltzmannian manner (see Truesdell, *Fundamentals of Maxwell's*, p. 171). For explanatory purposes, however, we need some way of motivating the belief that we may treat the positions and momenta as uncorrelated parameters. Therefore, it seems that Boltzmann's method and many other "intermediate positions" are either (i) specialized ways of assigning a priori probabilities; or (ii) they are imprecise approximations for a real proof, in the spirit of the ergodic program, of the overwhelming likelihood of an equilibrium; or (iii) they rely on a pragmatist argument in the spirit of Gibbs. We fail to see how a forth possibility may be characterized. Boltzmann, of course, was inclined to find dynamic explanations for the trend toward equilibrium, and in some of his writings we find the beginning of the ergodic approach as we now formulate it. For a thorough article on Boltzmann's relations to the ergodic approach, see von Plato, "Boltzmann's Ergodic Hypothesis," *Arch. Hist. Exact Sci.* (1991).

beginning of nonequilibrium thermodynamics. In this context, questions concerning the explanation of irreversible behavior are discussed in a setting that does not presuppose a notion of equilibrium. This important topic will not be discussed much in this book.[85]

Boltzmann's proof of the H-theorem involves very keen observations on the mechanical details of the collision process. It almost seems that the proof explains how disorder is "transported" throughout the system. Nevertheless, the H-theorem is statistical in character. *The theorem is an investigation of the effect of the collisions on an initial distribution and not on an individual system.* This rather refined point was not fully understood immediately, not even by Boltzmann himself. As a result, arguments raged for many years on the precise content and validity of the H-theorem. It seems that Boltzmann developed much of his conception of statistical physics while answering the objections of his critics. In particular, Boltzmann was preoccupied for a long time with the arguments of Loschmidt and Zermelo.[86]

Boltzmann began his 1872 article with a definition of an equation that expresses the change of the velocity distribution in time.[87] More precisely, he used this equation to calculate the changes in the velocity distribution if one assumes that the velocities of the colliding molecules are uncorrelated.[88] Next, Boltzmann observed that if $f(x,t)$ is the number of molecules with kinetic energy x at time t, then $\partial f/\partial t$ equals the number of molecules whose velocities did not change plus the number of molecules whose velocities became x at $+\Delta t$ minus the number of molecules whose velocities ceased to be x at t.

In other words,

$$f(x,t+\Delta t)dx = f(x,t)dx + \int dv - \int dn$$

where

$$\int dn = t\,dx \int_0^{00} \int_0^{x+x'} f(x,t)f(x',t')f(x,x',x)dx'\,dx$$

and

---

[85] For an interesting and informative account of the various issues involved in the nonequilibrium theory, see Sklar, *Physics and Chance*, Cambridge Univ. Press (Cambridge, UK, 1993), pp. 196–296.

[86] See Brush, *The Kind of Motion*, pp. 2442–248.

[87] Later formulations of the theorem describe the change of the energy distribution in time.

[88] As you may recall, this is essentially Maxwell's assumption.

$$\int dv = t\,dx \int_0^x \int_{x-u}^{00} f(u,t)f(x,t)f(u,v,x)\,du\,dv$$

$$+ t\,dv \int_x^{00} \int_0^{00} f(x,t)f(v,t)f(x,x',x)\,du\,dv.$$

u and v are taken as the energies of two molecules before a collision that results in one of them acquiring energy x afterwards. (The calculation uses the principle of integrals conservation of energy). Boltzmann's next step was to choose new variables, introduce some symmetry requirements on f, and exchange the order of the integration. As a result, Boltzmann finally obtained his transport equation:

$$\partial f(x,t)/\partial t = \int_0^{00} \int_0^{x+x'} \left[f(x,t)/\sqrt{x^x} f(x+x'-x,t) \middle/ \sqrt{(x+x'-x)} \right.$$
$$\left. - f(x,t)/\sqrt{x^x} f(x',t) \middle/ \sqrt{x'} \right] (x,x')^{1/2} f(x,x',x)\,dx\,dx.$$

If we substitute Maxwell's distribution and set

$$f(x,t) = C\sqrt{x}e^{-hx}$$

(where, as usual, C and h are constants), the factor in the square brackets vanishes and, as a result,

$$\partial(x,t)/\partial t = 0.$$

In other words, we have a proof that the Maxwell distribution does not change in time. That was the first result obtained by Boltzmann in the article. Next Boltzmann defined the H-function:[89]

$$H = \int f(x,t)\{\log[f(x,t)/Vx] - 1\}\,dx.$$

Boltzmann investigated the time derivative of H and proved that

$$dH/dt < 0$$

with equality only if f is the Maxwell distribution. The next argument that Boltzmann developed was designed to show that the quantity $-H$ has the same properties as the entropy. Therefore, Boltzmann con-

---

[89] Boltzmann originally used the letter E. But later, following the notation introduced by others, he himself started referring to the quantity as H.

cluded, the definition of entropy can be extended to nonequilibrium states.[90]

Boltzmann's transport equation has been used in a wide variety of applications ever since its initial appearance. Predictions based on this equation have been completely confirmed by empirical studies. The equation also supplies us with a very good theoretical connection between the microscopic and macroscopic levels of description. Nevertheless, these success stories only dramatize the strange character of Boltzmann's theory in particular and of the statistical study of mechanical systems in general. The results necessitate statistical assumptions of various kinds, but physicists do not have very good justifications of these assumptions. Indeed, what is the theoretical justification of the Stosszahlansatz?

1.9. THE CANONICAL DISTRIBUTION OBTAINED BY THE USE OF A VARIATIONAL PRINCIPLE

As we remarked earlier, Jaynes regarded the main questions of SM as inferential rather than dynamical. Therefore, it is not surprising that he believed that the standard dynamically oriented discussion on Boltzmann's H-theorem is fraught with confusion. For Jaynes, the only question worth asking was, why are we justified in predicting a trend toward disorder relative to any state of information? This formulation, however, is not compatible with the ordinary conception of the H-theorem. The transport equation is formulated as a mechanical theorem and not an inferential one. It is not at all obvious that the H-theorem can be recast in inferential terms Nevertheless, in SMIT Jaynes attempted to formulate and prove an inferential analog to the H-theorem.[91] He argued that if we think of the equations of motion H as having the form

$$H = H_0 + V_\alpha,$$

where H_0 is an "unperturbed" Hamiltonian and the V_α are external perturbations that happen at various points in time, then the increase of entropy merely expresses our decreasing ability to predict the future

---

[90] The discussion in this section follows rather closely Brush, *The Kind of Motion We Call Heat*, pp. 443–447.

[91] See ibid, pp. 27–28. The proof that Jaynes gave to his subjective H-theorem is in the appendix of SMIT II, ibid, p. 38.

states of the system. This argument is rather unsatisfactory. One reason is that, at least in classical contexts, one may attempt to incorporate the mechanism that generates the perturbations into a more general unperturbed Hamiltonian H'.[92] Relative to H', there are no "external disturbances" and hence there is no prima facia reason why our ability to predict should diminish. Jaynes, too, must have been dissatisfied with this argument because, in his later writings, he found a different way of demonstrating that the increase of entropy is compatible with his subjectivist philosophy. The idea was to make use of Gibbs's definition of entropy rather than Boltzmann's. Gibbs's conception, as we shall see, is more amenable to a subjectivist treatment.

Let M be a system that is observed at time t to be in a state of equilibrium with its surrounding. Such a system will not be completely isolated, but we do not expect large deviations around its mean energy E. Let $E_1, \ldots, E_n$ be the possible energy levels of M.[93] Let us then attempt to choose a probabilistic assignment $p(E_1), p(E_2), \ldots, p(E_n)$ to the events that M occupies, the energy levels $E_1, \ldots, E_n$, respectively. As you may recall, the assignment has to have the following properties:

1. $\sum_{i=1}^{n} p(E_i) = 1$

2. $\sum_{i=1}^{n} p(E_i) E_i = E$.

Suppose that we accepted the MEP. In such a case we could use a variational principle and insist that we select from among all of the assignments that satisfy properties 1 and 2 the one that maximizes the expression

$$-\sum_{i=1}^{n} p(E_i) \log p(E_i). \qquad (*)$$

It has been known for a long time that there is an assignment that satisfies the two constraints and equation (*). This assignment is Gibbs's canonical distribution. This fact allowed Jaynes to conclude

---

[92] If, on the other hand, one resists the idea that the perturbations are generated by a mechanism, there are no grounds for expecting that there will always be more perturbations. Consequently, one may not necessarily expect that entropy will always increase.

[93] The number of energy levels may be infinite. The argument in such a case will remain essentially the same.

that the canonical distribution is simply the MEP solution in the situation described above.

Now, to say that the canonical distribution minimizes (*) means that

$$\sum_{i=1}^{n} p(E_i) \log p(E_i) < \sum_{i=1}^{n} p'(E_i) \log p'(E_i)$$

as long as p' satisfies properties 1 and 2. Therefore, if for some reason the distribution over the energy levels $E_1, \ldots, E_n$ changes in time, the new entropy cannot increase. (Observe that all of the future distributions of M will continue to obey properties 1 and 2.[94]) Moreover, as Tolman observed,[95] when systems are in equilibrium with their environment, the distribution of the energy levels will not change very often. Therefore, taken together, the two insights give rise to a particularly simple version of the H-theorem. First, the entropy of systems that are in a state of equilibrium with their surroundings will not increase. Second, such systems will not stray from their maximally entropic state very often. Jaynes regarded this argument as a demonstration of the second law of thermodynamics.[96]

1.10. JAYNES AND GIBBS

Gibbs's version of the H-theorem does not give any mechanical account of the process whereby a single mechanical attains equilibrium. Gibbs was aware of this point, and he tried to supplement his statistical account with a suggestive but imprecise description of the process through which a single system becomes maximally disordered. The example that he used was the mixing of a drop of ink in a glass of water.[97] The ink and water system will reach equilibrium when equal parts of water will have similar ink content. This equilibrium state is stable; further mixing will not separate the ink from the water. The discussion on the process of mixing is conducted in the twelfth chapter of *Elementary Principles*, a chapter that Jaynes regarded as "confounded and confused," and the work of a terminally ill person.[98] This is, by any account, an extreme judgment. The twelfth chapter is clearly a work of

---

[94] For a discussion on Gibbs's H-theorem, which obviously influenced Jaynes, see Tolman, ibid, pp. 170–174.
[95] See "Principles of SM," ibid.
[96] See Jaynes, "Gibbs vs. Boltzmann Entropies," ibid, pp. 77–87.
[97] See Gibbs, *Elementary Principles*, ibid, pp. 144–145.
[98] See Jaynes, "Delaware Lectures," ibid, p. 100.

remarkable brilliance. It took mathematicians 30 years to translate Gibbs's prose into symbols, and when they finally did, the results did not improve much on Gibbs's treatment as an illumination of the physical aspects of the problem. Nevertheless, the place of the twelfth chapter is not clear. The remarks in that chapter seem to contradict the first chapters of *Elementary Principles*, where Gibbs called into question not only the necessity of proving ergodicity, but also the necessity of investigating the dynamic behavior of individual systems in the context of SM. Gibbs wrote:

> The usual point of view, in the study of mechanics, is that, where the attention is mainly directed to the changes which take place in the course of time in a given system. [. . .] For some purposes, however, it is desirable to take a broader view of the subject. We may imagine a great number of systems of the same nature, but differing in the configurations and velocities which they have at a given instant. [. . .] And we may set the problem not to follow a particular system through its succession of configurations, but to determine how the whole number of systems will be distributed among the various conceivable combinations of configurations and velocities at any required time when the distribution has been given for some one time. The fundamental equation of this inquiry is that which gives the rate of change of the number of systems which fall within any infinitesimal limits of configuration and velocity.[99]

As we remarked before, Gibbs's main conceptual breakthrough was his new definition of equilibrium. According to this definition, an ensemble is in equilibrium relative to a given distribution if the distribution does not change in time. This definition summarizes all of the dynamical issues that are relevant to inferential contexts in terms of this simple condition. The second important step that Gibbs made is the introduction of the canonical distribution, which is the simplest case of a distribution that remains constant in time. This distribution, which "mimics" the concept of temperature, is obtained when the logarithm of the probability is a linear function of the energy.[100] Now, canonical distributions have the following remarkable property. When we restrict

---

[99] See Gibbs, *Elementary Principles*, p. vii.
[100] Gibbs introduced the canonical distribution in the following passage: "The distribution represented by
$$n = \log P = f - e/q$$
or
$$P = e^{f-e/q},$$
where f and q are constants and e is positive, seems to represent the simplest case

our attention to subensembles, we find that they are very similar statistically to the original ensemble. Because of this property, when canonical ensembles are concerned, we may disregard any particular information that we have concerning M as statistically irrelevant. Suppose, for example, that we learn that our system has some property A. Ordinarily, that information would be reflected in our inferences: We should make our predictions relative to the subensemble of systems that have property A. But when canonical ensembles are concerned, because the various properties are distributed in this subensemble as they were distributed in the original ensemble, the information that the system has the property A is statistically irrelevant. Because of this property, Gibbs reasoned, it has to be the case that the ensemble averages are the most probable values in the different subensembles as well. So Gibbs's rules advise us to predict that, as time goes by, the system will assume these average values relative to any state of information. As we mentioned earlier, another point emphasized by Gibbs is that in the thermodynamical limit, that is, when the number of particles approaches infinity, the averages relative to the microcanonical and the canonical distributions are practically the same. This point is often used to justify the method of discarding information that is relative to the microcanonical distribution as well. In particular, some writers use this fact as an answer to the two questions that we mentioned in the beginning of this section. We can simply ignore the information that the system was prepared not to be in equilibrium because the subensemble of such systems has the same statistical properties as the original ensemble. In other words, future equilibrium is just as probable relative to this subensemble as well. The same argument applies to the subensemble of systems that are presently in a state of equilibrium. Such systems are just as likely as the general population of systems to be in equilibrium in the future. As we said before, these questions constitute an inferential analog to the question of irreversibility. Therefore, the answer that we just gave can be taken as a nondynamic version of the H-theorem.

What, then, is the status of the twelfth chapter? More generally, what

conceivable, because it has the property that when the system consists of parts with separate energies, the laws of the distribution in phase of the separate parts are of the same nature, a property that enormously simplifies the discussion and is the foundation of extremely important relations to thermodynamics" (Gibbs, *Elementary Principles*, ibid, p. 33). In systems that are in thermal equilibrium, the different parts are internally in a state of thermal equilibrium.

is the status of Gibbs's ensembles? Since 1902, when Gibbs wrote his *Elementary Principles*, there have been constant attempts by physicists to understand the precise status of Gibbs's ensembles and why they yield the correct experimental results. Some (the Ehrenfests, for example[101]) concluded that Gibbs's rules are too "sketchy" and ad hoc. Others actually found ways to reformulate Gibbs's ideas rigorously and provide them with partial proofs. One of the most successful attempts in this direction involves an argument based on the fact that the microcanonical distribution is sharply peaked around its average and has a small dispersion around the average value. This is a consequence of the Khinchin–Lanford theorem. Because of the single sharp peak around the mean, if we are to predict the state that the system will assume, the mean will be the best prediction. Also, because the dispersion is small, we may conclude that the system will not "stray too far." These points are what constitute the "inferential analogy" to the issue of irreversibility.[102] But the main problem with this argument is that it leaves the most important question unanswered. What we need to know is the physical significance of the canonical *or* the microcanonical distribution. Because of this problem, Jaynes's point of view is rather compelling. He asked: Why shouldn't we admit that SM probabilities reflect an epistemic attitude instead of physical properties?

Jaynes, however, did not argue that his epistemic interpretation should be accepted by default. In a passage that should be regarded as one of Jaynes's most important contributions to the discussion of the *H*-theorem, we find a positive reason for his view.[103] "In QM as in classical theory," he wrote, "the increase in entropy characteristic of irreversibility always signifies and is identical with a loss of information. It is important to realize that the tendency of the entropy to increase is not a consequence of the laws of physics as

---

[101] See P. and T. Ehrenfest, *The Conceptual Foundations of the Statistical Approach in Mechanics*, Cornell Univ. Press (Ithaca, NY, 1959): "However, as soon as we come to the main object of Gibbs' book, i.e., to the nonstationary ensembles, we meet with rather sketchy arguments. The reason for this is that only here do we have to begin to take into account the dynamical character of the statistical ensemble in question" (p. 52). For a summery of the Ehrenfests' critique of Gibb, see also Sklar, *Physics and Chance*, ibid, pp. 67–71.

[102] For a summary of the relevant results on the Khinchin–Lanford theorem (Khinchin 1948, Lanford 1965), see Malament and Zabell, "Why do," ibid. It remains unclear, though, how the authors suggest treating the stochastic behavior of systems with a small number of degrees of freedom.

[103] See Jaynes, ibid, *Papers*, p. 26.

such. [. . .]"[104] This loss of information, Jaynes says, may not even be remotely connected to any physical reason. "An entropy may increase unavoidably," he writes, "as a result of our incomplete knowledge of the system, or it may be an entirely voluntary act on our part. In the latter case, an entropy increase is the means by which we simplify a prediction problem by discarding parts of the available information which are irrelevant or nearly so, for the particular prediction desired."[105] The importance of this remark is clear. Given that the entropy increase can be introduced in an artificial way, the use of Gibbs's predictive techniques cannot always be given an objective physical justification. It is also clear, however, that the idea of discarding voluntarily "irrelevant or nearly irrelevant information" is rather controversial. If, by irrelevance, Jaynes means probabilistic irrelevance relative to our subjective probabilities, we should not be expected to use Gibbs's rules too often. According to most Bayesians, information concerning the way in which the system was prepared should be considered, mutatis mutandis, as relevant. (In other words, if, relative to some probability distribution, such information is always irrelevant, so much the worse for using this distribution.) Therefore, even if, strictly speaking, the use of Gibbs's rules is consistent with Bayesianism, it cannot be given a very natural interpretation in the standard Bayesian framework. On the other hand, if Jaynes is using irrelevance in some unspecified physical sense, he must be implicitly relying on an argument in the spirit of the ergodic program that attempts to establish the objective grounds for the view that in general we may be permitted to discard information concerning the mode of preparation of the system. We shall return to this dilemma later.

1.11. A CRITIQUE OF JAYNES'S THEORY

As we remarked earlier, we do not believe that Jaynes's theory constitutes a workable version of statistical mechanics. More generally, we do not believe in ultrasubjectivism. Probabilistic reasoning is not autonomous, and in different contexts we are bound to constrain our personal probabilities in different ways. These constraints – the different invariance principles – are what characterize the different contexts for the subjectivists. And there is no reason to expect there to be

---

[104] Ibid.   [105] Ibid.

any simple way to unify them. (At any rate, a unified presentation of the various invariance principles is a nontrivial mathematical project.). But these remarks do not amount to a serious criticism of Jaynes's theory, because we still have to show that the ultrasubjectivist version of SM yields a loss of information or explanatory power. Let us, then, finish this chapter by hinting at three different directions that such criticism may follow.

The first direction is quite straightforward. For Jaynes, the only reason that thermodynamic equilibrium exists is because the systems under consideration have a large number of degrees of freedom. That is why we can rely on the maximum entropy inference. When systems with few degrees of freedom are considered, the canonical and microcanonical distributions simply do not coincide. Nevertheless, there are reasons to believe that even systems that do not have a very large number of degrees of freedom may exhibit a uniform behavior. If this is indeed the case, a dynamic investigation of such systems seems like a promising idea. And, indeed, in the last decade or so many physicists have begun to take seriously the idea that simple deterministic systems could simulate complex stochastic behavior.[106] The writings on this subject and particularly the popular literature have been somewhat confused as to the reasons why this may happen. But the data strongly suggest that a simplistic approach to the concept of complexity may lead us to the wrong conclusions.[107] These results seem to us to bear directly on the predictive approach to SM in general and Jaynes's theory in particular. They suggest that the large number of degrees of freedom is not a necessary condition for statistical regularity. At best, it is an epiphenomenon or a sufficient condition. We may envisage then a more general theory that could give a uniform explanation for the statistical regularity of the behavior of systems of varying levels of complexity. More generally, one of the interesting issues to be raised concerns the strange overdetermination in explaining SM phenomena. In

---

[106] This conjecture is discussed in the literature on chaos theory. For articles on the subject see Mira, *Chaotic Dynamics*, World Scientific (1987) and Gluckheimer et al., *Dynamic Systems*, CIME lectures (Bressanone, Italy, 1978). See in particular the article by J. Moser, ibid. For a popular article see Crutchfield et al., "Chaos" by *Scientific American* (Dec. 1986). These ideas go back to Maxwell's suggestion that what introduces the statistical element into SM is not the large number of molecules but the collisions with the walls of the container. See Maxwell's article in *Trans. Cambridge Phil. Soc.* 12(547) (1879).

[107] See Crutchfield et al., "Chaos," ibid, p. 55.

many cases, both the large number of degrees of freedom and the particular character of the dynamics can explain the statistical regularity of SM systems. This overdetermination should sway people from taking dogmatic positions with regard to the identity of the "true cause" of the approach to equilibrium.

The second hint of ultrasubjectivism's weaker side is more substantive, although we are not in a position to say much about it. The point concerns Jaynes's view about the "anthropomorphic" nature of entropy, as he puts it.[108] Rather recently, it has been discovered that there are certain pairs of fluids that behave in the following way: After they have been mixed well initially, these fluids will separate from each other "spontaneously" when the temperature is lowered beyond a certain point. When the temperature is lowered even further, these fluids will begin mixing again![109] The explanation for this behavior is rather complex, and we shall not attempt to summarize it here. However, one key factor in the explanation is that the fluid system, in its "attempt to lower the entropy" (as the temperature is lowered), may either decrease the compositional entropy (by separating the fluids from one another) or decrease the directional entropy or the uniform organization of the molecules. Explaining when compositional entropy is lowered and when the directional entropy is lowered involves an argument that establishes the "optimality" of the strategy that the system "chooses." If such an explanation is a successful one, it points out that there cannot be any reference to an observer in the definition of entropy because it is the system and not the observer that "selects" the course of events. Thus, high entropy cannot be conceived of merely as an anthropocentric concept; that calls for an evaluation or even an empirical confirmation of the MEP. At any rate, the argument points to the possibility that the concept of entropy was misused by Jaynes.

The third line of criticism, however, is the most telling. We have alluded to it obliquely throughout this chapter. Ever since Jaynes published SMIT, there has been a steady flow of interesting articles written by philosophers and statisticians whose aim is to find flaws in Jaynes's claim that the MEP is a valid principle of inductive inference. These

---

[108] For a discussion by Jaynes on the anthropomorphic character of entropy, see, e.g., *Papers*, ibid, p. 85.

[109] For a popular presentation, see J. S. Walker and C. A. Vause, "Reappearing Phases," *Sc. Am.* (May 1987).

arguments are certainly relevant responses to Jaynes's claims about the a priori character of MEP. Indeed, the arguments of these writers cast a serious doubt on the thesis that we should always be guided by the MEP. However, in the context of SM there is no real doubt that use of the MEP will lead us to predict correctly. Jaynes's algorithm is, after all, a reformulation of Gibbs's rules, and these rules have been proven to be enormously successful. Therefore, the relevance of these arguments to the context of SM is not clear at first blush. In this context, as we have seen, the main problem is to explain why Gibbs's rules "work."

There are, however, at least two such arguments that have important consequences for SM. The first argument, in essence, is that unless Jaynes is willing to assume that the physical background is sufficient to determine the partition of the event space uniquely, the application of the MEP is sensitive to the choice of the partition. The reasons for the sensitivity to the choice of the partition are roughly the same as the reasons that make the consistent application of the law of insufficient reason so problematic. Relative to different partitions, the same events will be assigned different probabilities. The assumption that there is a privileged partition that is determined by the physical theory is, of course, a nontrivial objectivist assumption. So Jaynes's theory is either less subjectivist than he believes, or it introduces a partition sensitivity into SM.

The second argument is based on a result proven by Shimony and Friedman (1971, 1973)[110] and was extended to a philosophical critique by Seidenfeld (1979, 1983).[111] The argument can be explained with the following example. Let $P_1, \ldots, P_6$ be a space of events that represent the possible results of the throw of a die. P_i stands for the event that the throw yielded the number i on top. The distribution on $P_1, \ldots, P_6$ that obeys the MEP is the uniform distribution p; that is, for all i, $p(P_i) = 1/6$. Observe that in this case the expectation satisfies the following condition:

$$1 \times p(P_1) + \cdots + 6 \times p(P_6) = 3.5 \qquad (*)$$

Consequently, the uniform distribution is the MEP solution *among all of the distributions that obey* (*). Now, suppose that we obtain new information e that an odd-numbered side resulted on the roll. That

[110] See Friedman and Shimony, ibid.
[111] See Seidenfeld, "Why Am I Not," ibid.

information will surely persuade us to discard p for a new probability p^*, which concentrates on P_1, P_3, P_5. That is,

$$p^*(P_1) + p^*(P_3) + p^*(P_5) = 1.$$

Now, for Bayesians, p^* is simply the conditional measure $p(\ /e)$ that was obtained from p through conditionalization. It is easy to see that $p(\ /e)$ is uniform over P_1, P_3, P_5. That is

$$p^*(P_1) = p^*(P_3) = p^*(P_5) = p(P_1/e) = p(P_3/e) = p(P_5/e) = 1/3.$$

However, Jaynes does not give us very precise instructions as to how to proceed in this case. On the one hand, his position suggests that we apply the MEP to the new space of events and simply forget about the fact that our old assignments were constrained by the equation (*). This will result in a new distribution p^*, which will agree with the Bayesian solution. A second option, though, is that we treat the new information e as a constraint that we add to (*) and relative to which we calculate the MEP solution. If we followed the second option, the result would be the following:[112]

$$p^*(P_1) = 0.21624$$

$$p^*(P_3) = 0.31752$$

$$p^*(P_5) = 0.46624.$$

What, then, should a rational agent choose to do? It seems that there is an indeterminacy in Jaynes's prescriptions and a potential conflict between the Bayesian and the Jaynesian strategies for updating one's degrees of belief in the face of new information.

Note that this conflict will not arise when the new information leads us to a new probability distribution p^* that concentrates on a set of events with p-probability zero or p-probability one.[113] In this case, both recommendations coincide. This point is very important for us because, in the context of SM, we are usually confronted only with such cases. When we receive information in a routine way, Gibbs's rules tell us to ignore it. Such a case can be construed as $p^* = p(\ /e) = p$, which could be the case if $p(e) = 1$. The only other case is one in which we obtain new information that lowers the dimension of the space of serious

---

[112] See Seidenfeld, "Why Am I Not," p. 473 and his appendix A.
[113] This is a consequence of Seidenfeld's result.

possibilities, for example, when we learn about the energy level of the system in question. In such cases, p^* concentrates on a set with p-probability zero, and there is no conflict between conditionalization and the MEP. However, the reason why we do not expect nontrivial conditionalizations in the context of SM is precisely what the ergodic approach tries to explain. Nonergodic systems have nontrivial invariant parts P_i such that $0 < m(P_i) < 1$, which means that, if the system is in $s \ \varepsilon \ P_i$ at one point, it will stay there forever. Such information is clearly very relevant for the purposes of prediction. Such information also will lead us to either use the MEP relative to the knowledge that the system is at P_i or adopt a nontrivial $p^* = p(\ /s - P_i)$ through conditionalization on the knowledge that the system occupies a state in P_i. In such a case, all of Seidenfeld's worries will materialize, namely, use of the MEP will contradict the principle of conditionalization.[114]

Notice that those who accept the ergodic approach may try to resolve the conflict by stating that the MEP is justified in the context of SM because there are objective physical reasons as to why nonergodic systems are not likely to be found. However, because Jaynes objects to the ergodic approach, what makes it reasonable for him to suppose that we shall never encounter nonergodic systems? In other words, the reliance on the Bayesian framework does not constitute an argument against the ergodic approach. On the contrary, it is rather obvious that, without a proof of ergodicity, the use of Gibbs's rules is something of a mystery from a Bayesian point of view.

---

[114] The reader should note that the current discussion on the status of the MEP principle is lively and complex, and many issues connected to the discussion were not mentioned here. For a taste of this discussion, see, for excample, Skyrms's investigation of the relations between conditionalization and the MEP. [See Skyrms, "Maximum Entropy Inference as a Special Case of Conditionalization," *Synthese*, 63(1) (1985). In fact, the whole volume is concerned with the MEP.] Note that Skyrms's discussion contains the following ingredients: (i) There is an emphasis on the fact that the discussion on the relations between condition-alization and the MEP necessitates an extension of the "space of events." (ii) Skyrms's solution uses the notion of second-order probabilities. Both of these ideas are difficult to interpret in the framework of the discussion on SM. The reader should also note that Jaynes made some attempts to answer his critics. [See, for example, Rosenkrantz, "Where Do We Stand on Maximum Entropy?," in *Papers*, ibid, and Jaynes, "Some Random Observations," *Synthese* 63(1) (1985).] We do not believe, though, that Jaynes offered a solution to the main problem.

1.12. SUBJECTIVISM VERSUS PRAGMATISM IN THE FOUNDATIONS OF SM

The aim of this chapter has been to present Jaynes's view in an ultra-subjectivist framework. According to this presentation, Jaynes's aim was to "strip down" SM and reconstruct it as a straightforward application of the rules of subjectivist statistical inference. We saw that Jaynes's strategy for achieving this goal was to rid SM of "all but purely mechanical laws," and to concentrate on inferential rather than mechanical issues.[115]

Some aspects of this program were proven rather successful. Using his MEP rule (which we regard as an optional but rather natural symmetry constraint), Jaynes derived both the canonical and the microcanonical distributions. Thus, a person who does not have any information concerning the mode of preparation of a many-particle system M and who agrees to use the MEP will predict that M will most probably assume states that are compatible with macroscopic equilibrium. This derivation relies solely on a priori considerations. However, when we are not completely ignorant, our ability to predict the "correct" future of M depends on the availability of a reason for discarding the additional information that we possess. Gibbs showed how relying on his canonical distribution furnishes us with such a reason. But what could constitute an argument for the use of the canonical distribution? If Jaynes forbids us to look for an objective reason, we have to fall back on the standard Bayesian reasoning, but we have argued that such reasoning is not very conducive to an acceptance of Gibbs's rules. The dissonance is a deep one, because the very idea of discarding available information goes against the grain of the inductive method.

There is, however, a very different undercurrent in Jaynes's writing that has little to do with ultrasubjectivism. Sometimes Jaynes seems very close in spirit to Gibbs's pragmatist approach. Take the following passage, for example, in which Jaynes compares the status of SM with QM:[116]

The unceasing confusion that swirls about the Copenhagen interpretation of quantum theory is, in my opinion, the direct result of a very similar but more subtle, misplaced emphasis. The mathematical rules of present quantum

---

[115] Jaynes, *Papers*, ibid, p. 87.   [116] Ibid.

mechanics, like Gibbs's rules, are highly successful and clearly contain a great deal of very fundamental truths. But nobody knows what they mean; they are in part expression of laws of Nature, in part expression of principles of human inference, and we have not yet learned how to disentangle them.

In this paragraph it seems that Jaynes argues that we should accept both QM and Gibbs's rules as true theories even if they contradict some of our assumptions about the dynamics of physical systems. In both cases, the value of the theories is that their predictions agree with the experimental results. But to explain their success we have to relinquish, respectively, the hidden variable interpretation in the case of QM and the search for consistent dynamic explanations in the case of classical SM. This plea seems very much like a pragmatist justification scheme.

In his first article, Jaynes was not so extreme. In a section of SMIT I, entitled "Subjective and Objective SM," he wrote:

Although the MEP appears capable of handling most of the prediction problems of SM, it is to be noted that prediction is only one of the functions of SM. Equally important is the problem of interpretation; given a certain observed behavior of the system, what conclusion can we draw as to the microscopic causes of that behavior? To treat this problem and others like it, a different theory which we can call objective SM is needed. Considerable semantic confusion has resulted from failure to distinguish between prediction and interpretation problems, and attempting a single formalism to do both.[117]

Jaynes never offered an objective SM that was suitable for explanatory purposes. Moreover, some of his later remarks suggest that he might have reconsidered the view that an objective SM was, indeed, necessary. Following the example of QM, he might have reverted to a view that stresses the predictive rather than the explanatory aspects of physics in general. If Gibbs's rules are "in part expressions of laws of nature; in part expressions of human inference and we have not yet learned how to disentangle them," it is doubtful whether we can really attempt an explanation of "certain observed behavior" in terms of "microscopic causes of that behavior." The most we can hope for is to "derive" the correct observed behavior.

The shift from subjectivism to pragmatism is rather familiar in the context of QM. But we should be aware of the fact that in the context

---

[117] Jaynes, p. 13.

of SM, too, these two tendencies coexist side by side and should not be confused. Moreover, we know that both the MEP and the idea of discarding information to simplify a prediction problem can seem more convincing when stated in the context of a pragmatic justification of SM.[118] In the last chapter of this book we shall return to the discussion on the pragmatist foundations of statistical mechanics.[119]

---

[118] For another fairly pragmatist quote from Jaynes's writings, see ibid, p. 416. "Our aim is not to 'explain irreversibility' but to describe and predict the observable facts. If one succeeds in doing this correctly, from first principles, he will find that philosophical questions about the 'nature of irreversibility' will either have been answered automatically, or else will be seen as ill considered and irrelevant."

[119] Note that in this chapter we do not pretend to give an argument against the use of the H-theorem approach. We are merely suggesting that the ultrasubjectivist interpretation of the H-theorem approach that we ascribe to Jaynes fails to provide a coherent interpretation of SM.

2

Subjectivism and the Ergodic Approach

2.1. THE ATOMISTIC CONCEPTION IN LIGHT OF THE RESEARCH ON THE BROWNIAN MOTION

In 1909, William Ostwald, formerly one of the most vocal opponents of atomism, admitted that, "recently we have come into possession of experimental proof of the discrete or granular nature of matter, for which the atomic hypothesis had vainly sought for centuries, even millennia."[1] The proof consisted of the "isolation and counting of gas ions" and "the agreement of Brownian movements with the predictions of the kinetic hypothesis."[2] Ostwald commended J. J. Thomson and J. Perrin for the former and later parts of the experimental work, respectively.

The interest in atomism and the willingness to consider adopting it as a new theory of matter was not confined to the community of physicists. Due to the efforts of Perrin, the topic was discussed and debated in a wide range of semipopular publications and lectures. The result was a fairly wide-ranging acceptance of atomism, a change of world view that deserves to be called a conceptual revolution. On first blush, the willingness of the educated world to reverse its opinion on the issue of atomism seems rather inexplicable. Only nine years earlier, Boltzmann was practically the only physicist on the continent who espoused the atomistic conception of matter.[3] One should remember, though, that the climate of opinion at the time, especially on the continent, favored new and extreme views.[4] Indeed,

---

[1] See Ostwald, *Grundriss der allgemeine Chemie*, Engelmann (Leipzig, 1909).
[2] Ibid.
[3] See Brush, *The Kind of Motion We Call Heat*, North Holland Publishing (Dordrecht, 1976).
[4] For a book about the climate of opinions at the end of the last century, see H. S. Hughes, *Consciousness and Society*, Knopf (New York, 1958).

both Mach's radical phenomenalism and Perrin's strange new world of "real" atoms exerted considerable fascination at the time. Another fact worth mentioning is that the recent invention of the microscope might have helped to convince people to broaden their conception of what is "immediately given to the senses," or, at the very least, it might have helped people to realize that the distinction between the real and the theoretical was not a simple matter.

Perrin argued for atomism in two different ways.[5] First, he attempted to give an indirect experimental proof for the existence of molecules. He did this by showing that the strange erratic motion of relatively small particles that are suspended in liquid, a phenomenon known as the Brownian motion, is the result of the bombardment of scores of invisible particles. The various impacts of the molecules, which come from different directions, are not expected to result in a lasting displacement of the Brownian particle. But, in the short run, fluctuations of the positions of the suspended particles may be observed, resulting in a Brownian motion. This argument, as simple as it may sound, was not well understood before the papers of von Smulochowsky and Einstein on the subject.[6] Earlier writers noted that a single molecule could not produce any visible impact on the Brownian particle. They did not, however, conceive of the possibility that the visible motion of the particle in one direction is the sum of the impacts in various different directions. The different impacts are expected to sum up to zero in the infinitely long run. But the temporary fluctuations of the partial sums from zero may produce a net impact that may be visible through a microscope.

Perrin's second argument was directed against an objection to the statistical-atomistic approach, which received a lot of attention in the 1890s. The objection, whose main proponent was Zermelo, was the following: One of the consequences of the second law of thermodynamics is the irreversible trend toward a maximally disordered state. This consequence has been completely verified experimentally. From the point of view of the statistical-atomistic approach, though, this irreversible trend is rather puzzling. The mechanical motions of the atoms are presumed reversible; therefore, where does the irreversibility come from? It seems that there is an essential tension between

---

[5] See *The Kind of Motion We Call Heat*, pp. 655–705.
[6] See Einstein's article in Ann. Phys. 4(17), (1905), p. 549 and von Smoluchovsky, *Rozprawy Krakow A* 46 (1906), p. 257.

the reversible dynamics of the molecules and the irreversible dynamics of the second law of thermodynamics. But not every tension amounts to a contradiction. Boltzmann solved the puzzle when he argued that the irreversible trend toward disorder, while being the most probable course of events, is not without exceptions. Yet, one aspect of the puzzle remained troubling. If the second law of thermodynamics has only a statistical validity, why is it that we never observe any exceptions? Why don't we ever observe a reversal of the trend toward disorder? This puzzle convinced Zermelo, among many others, to abandon the statistical-atomistic conception altogether. It is worth mentioning that his argument also was rather well publicized. The idea of a global irreversible trend toward disorder, to which popular writers referred by the term "the entropic death of the universe," resonated with the general fascination with doom and apocalypse. There are numerous references to the second law of thermodynamics in the popular literature and journalism from the end of the last century.[7]

Perrin's answer to Zermelo was as follows. The Brownian phenomenon proves that it is indeed the case that the second law is only highly probable but not completely certain, and fluctuations may indeed be observable. Moreover, he stated, one of the consequences of the theory of Brownian motion is that one may, at least in theory, convert "heat motion" to mechanical motion, thereby contradicting the second law. A demonstration of this possibility was given in terms of the following thought experiment. Suppose that we insert a membrane into a liquid. The membrane chosen is permeable to the particles of the liquid but not to the Brownian particles. Therefore, if we put all of the Brownian particles on one side of the membrane, an osmotic pressure will be created and the particles will bombard the membrane continuously on one side. Now suppose that we connect the membrane to a weight. If the particles are located above the membrane, when the membrane moves it will pull the weight with it. This, then, is a proof that one may convert the "heat motion" of the Brownian particles into work. Perrin, building on the work of Einstein and von Smoluchovsky, showed that

---

[7] Brush claims that the interest in the second law of thermodynamics and the concept of entropy was connected with the neo-romantic fascination with the idea of degeneration. See Brush, "Thermodynamics and History," *Graduate J.* VII(2) (Spring 1967).

the Brownian motion provides an experimental confirmation of one of the predictions of the statistical approach, namely, the possibility of visible fluctuations from the state of maximal disorder. In other words, he demonstrated that the second law has only a statistical validity.

These developments changed the methods and aims of statistical mechanics. In the nineteenth century, the idea that matter is composed of a large number of aimlessly moving molecules was invented by physicists whose research aim was to answer the most elementary questions concerning the mechanisms of diffusion, heat transfer, pressure, viscosity, and so on. Such questions could not have been answered by merely looking at the phenomenological equations that describe the relations between macroscopic parameters, such as pressure, volume, temperature, and so on. Equations such as Boyle's law simply cannot help us even to formulate an educated guess concerning, for example, the effect of gravity on a column of gas or the effects of viscosity. Therefore, physicists such as Maxwell, who wanted to answer these questions, had to "transcend the phenomena" and invent detailed mechanical models that helped them to formulate hypotheses that could be tested experimentally. A particularly important example of this kind is the way Maxwell used molecular collisions to explain the mechanism of heat transfer. Indeed, after the demise of the caloric theory of heat, whose proponents regarded heat as a substance, there was no other way to understand the mechanism of heat transfer. If heat is not a substance, people reasoned, how can it be transferred from one place to another? The explanation that Maxwell offered was given in terms of molecular collisions. The cooler body will have, on the average, slower molecules. Therefore, it is likely that when the particles of the two bodies collide, the energy transfer will proceed from the warmer body to the colder one until both bodies have the same average kinetic energy.

Maxwell's investigations are good examples of the character of early studies in kinetic theory. Using molecular models, these studies provided answers to questions that phenomenological investigations were incapable of answering. One may say then that, at this point in the development of physics, the construction of deep-level descriptions of matter was driven by the wish to answer nagging questions rather than by a desire for grand unifying schemes. This attitude is also characteristic of Boltzmann's investigations. Boltzmann's atomism enabled him to obtain such remarkable results as the equipartition theorem[8] and the transport equation.[9] But even Boltzmann, who wrote as if he believed in the reality of molecules, claimed that ultimately his molecular models were needed merely to facilitate his inquiries. And toward the end of his life he explicitly repudiated any form of noncritical realism concerning atoms.[10]

In summary, the attitude towards atomism in the late nineteenth century was tenuous and tentative. Insofar as the kinetic models were productive, physicists tolerated them and even encouraged their use; but when they yielded puzzles, contradictions, and paradoxes, the scientific community turned its back and repudiated them. Some writers, notably Lord Kelvin, Poincaré, Loschmidt, and Zermelo, reversed their opinions completely. As we mentioned earlier, the problems that bothered this group of writers were connected to the issue of irreversibility. Other writers, such as Gibbs, formed the view that the concepts of statistical mechanics stand in a weak analogical relation to the thermodynamic concepts that they were supposed to explicate. Boltzmann, instead, did not think that any of the difficulties was insurmountable, and he tried to solve them. He insisted that the phenomenological equations simply did not contain enough information. Therefore, he concluded, there is no way to proceed without mechanical models.

Another path that led Boltzmann to atomism was his finitism. He insisted that continuous mathematics had to be reformulated in terms of small and discrete quantities for it to be capable of describing physical phenomena properly.[11] This motivation, too, has little to do with any form of realism with respect to atomic particles; Boltzmann never argued that atoms were simply the "furniture of nature."

For Einstein, the situation was quite different. In a letter that he wrote to Mach in 1909, Einstein emphasized that the Brownian motion "is a motion that one *must* believe to be a heat motion"[12] (italics

---

[8] The theorem proves that the energy is equally divided, on the average, between all of the different degrees of freedom. In particular, in a mixture whose elements are particles with different masses, each element in the mixture has the same average kinetic energy. See Brush's *Statistical Physics*, pp. 65–76.

[9] See Chapter 1, Section 1.6.

[10] See Boltzmann's "Unterberlichkeit der Atomistik in der Naturwissenschaft" in his *Populaere Schriften*, Barth (Leipzig, 1905), pp. 145–157. See also his "Nochmals ueber die Atomistik," ibid, pp. 158–161.

[11] See ibid, p. 144.

[12] The letter is dated 9.8.1909. See *The Kind of Motion*, p. 700.

added). In other words, the belief in the kinetic model was, according to Einstein, no longer predicated on the utility of the model, that is, on its ability to facilitate the inquiry concerning the macroscopic properties of matter. The atomistic conception, which was soon to be given foundations in quantum physics, was regarded by Einstein and his contemporaries as the most fundamental physical theory. According to the modern point of view, atoms were not entities postulated merely to "save the phenomena." From this point on, nothing short of a scientific revolution could bring modern physicists to question the atomic theory of matter.

One of the consequences of the modern view of atomism was a reformulation of the aims and methods of statistical mechanics. According to the modern point of view, "the excellent agreement between the predictions of statistical mechanics and results of experiment [is not] in itself sufficient justification of the methods used in statistical mechanics . . . a satisfactory foundation of statistical mechanics has been sought whereby its methods may be seen to be extensions of exact mechanics classical or quantal."[13] This formulation certainly aspires to go beyond Gibbs's pragmatist attitude, but in truth it marks a departure from Boltzmann's methodology as well. In the nineteenth century statistical mechanics was required to facilitate the research in thermodynamics without creating contradictions. In other words, the most important foundational issue was whether SM is consistent with thermodynamics. But after 1905 physicists began to attempt to demonstrate that SM actually follows from purely mechanical principles. In other words, the issue of reduction was introduced into the discussion.[14]

In addition, there were clear differences of emphasis and intellectual style. The proponents of the new approach to statistical mechanics were versed in abstract mathematics, more convinced about the inherent importance of foundational studies, less oriented toward the discovery or explanation of new physical phenomena, and more interested in the mathematical and philosophical study of probability theory. This line of research is the topic of this chapter. More specifically, we shall discuss one of the most important

---

[13] See Farquhar, *Ergodic Theory in Statistical Mechanics*, Wiley and Sons (London and New York, 1964), p. 3.
[14] In the nineteenth century, the discussion revolved around the idea of reduction of properties and not reduction of theories.

directions taken by the modern research, namely, the ergodic approach.[15]

2.2. *THE CONCEPTUAL FOUNDATIONS OF THE STATISTICAL APPROACH TO MECHANICS* BY P. AND T. EHRENFEST

The modern conception of SM received its first definitive formulation in an encyclopedia article written by Paul and Tatiana Ehrenfest in 1912. The article was subsequently published as a book entitled *The Conceptual Foundations of the Statistical Approach to Mechanics.*[16] The book remains, to this day, one of the most influential contributions to the literature on SM. Of particular importance is the fact that, for the first time, the ergodic approach to SM was discussed at length. The Ehrenfests were fully conscience of the rift that separated the modern formulation of SM from the older one; in fact, the book is structured around this distinction. The first part contains a discussion of the main contribution to SM before the controversies of the 1880s and the 1890s. In the second part they discuss the more sophisticated statistical formulations that mark the modern treatment of the subject. Finally, in the third part they discuss, rather critically, the writings of Gibbs.[17]

Apart from the fact that *The Conceptual Foundations* discusses the various substantive modern developments in the field, the book embodies the methodological shift that characterizes the modern approach to the foundations of SM. For a book on physics, *The Conceptual Foundations* is remarkable in the painstaking attention that is paid to methodological issues, the lengthy discussions it contains on purely mathematical points, and the deep and informed treatment of various statistical topics. As we mentioned earlier, these are among the distinguishing marks of the modern discussions on the foundations of SM.

---

[15] We present this chapter as yet another alternative to Gibbs's pragmatist approach. Yet there are writers who see Gibbs as the father of the ergodic approach. The source of this confusion is the view expressed in the twelfth chapter of *Elementary Principles*, which we discussed in the first chapter.
[16] See P. and T. Ehrenfest, *The Conceptual Foundations of the Statistical Approach in Mechanics*, Cornell Univ. Press (Ithaca, NY, 1959).
[17] For a discussion on the Ehrenfests' critique of Gibbs, see Sklar, *Physics and Chance*, ibid, pp. 67–71.

Another point that deserves to be mentioned is that the Ehrenfests did not conceive of their article as an attempt to solve the foundational problems of SM. Originating as a survey article, the book emphasizes the unsolved problems in the field. The Ehrenfests proposed a list of questions that needed answers, and generation after generation of physicists have tried to find the answers to them. It seems fit, then, to begin with an enumeration of these questions. Once we are acquainted with the questions, we shall be able to assess the answers to the questions that were proposed by the adherents of the ergodic approach.

We have already mentioned that some physicists (notably Einstein) regarded physical probabilities as observables. This view necessitated a new physicalist definition of probability. The challenge is to shy away from any assumption that cannot be tested empirically. The inclusion of such assumptions would immediately call into question the physical meaning of probabilistic statements. Einstein tried to solve the problem by defining probabilities as time averages.[18] We shall discuss this idea at length shortly.

The second problem in the foundations of SM was the justification of the principle that equal areas in the phase space should be regarded as equally probable. Since the 1870s there was a genuine doubt concerning the validity of this principle. The puzzle concerned the issue of specific heats. The expectation values calculated on the basis of the assumption of equal probabilities yielded the wrong predictions. The solution was obtained by Planck in 1905, when he showed that the discreteness of the permissible energy levels led to the problem. As we mentioned earlier, Gibbs was particularly troubled by this problem. One should note, though, that the ultimate justification of the indifference principle, as it is referred to in the philosophical literature, remained an open issue even after a solution to the problem of specific heats was obtained.

The third problem is the Stosszahlansatz. More broadly, it pertains to the status of the various assumptions of molecular chaos that were used to obtain many results in SM. To begin with, these assumptions seem to need a theoretical and empirical justification. Such a justification was conspicuously absent in the beginning of the century. Even more crucial, though, is the question of the consistency of the Stosszahlansatz with the underlying mechanical laws. This question of

---

[18] The idea appeared even earlier in the writings of Maxwell, Boltzmann, and Gibbs.

consistency may be presented relative to a classical mechanical framework or a quantum mechanical one.

The fourth problem concerns the consequences of Poincaré's recurrence theorem or the "Wiederkehreinwand," as Zermelo referred to it. The theorem states that in the infinite long run it is overwhelmingly probable that mechanical systems will return arbitrarily close to their initial states. This theorem, which is extremely general and abstract, certainly applies in the context of SM. Nevertheless, it seems to contradict the second law of thermodynamics.

The fifth problem is Loschmidt's "Umkehreinwand." Loschmidt argued that, for every mechanical path where the entropy of the system increases, there exists a precisely opposite path where the entropy has to decrease. This is an immediate consequence of the reversibility of the equations of motion with respect to the direction of time. Again, though, this fact seems to contradict the second law of thermodynamics.[19]

The sixth problem is a generalization of the forth and the fifth ones. It concerns the question of whether it is possible, in general, to deduce thermodynamic laws from purely mechanical principles or, at least, to show that there is no inconsistency between the two theories. One may generalize the problem even further and ask whether all of the laws that express macroscopic regularities can be reduced to purely mechanical laws. Naturally, the notion of reduction has to be defined properly before a solution to the question can be sought.

The seventh problem we shall mention concerns a mathematical issue. The earliest formulation of the idea that probabilities were definable as time averages invoked the assumption that, in due course, a mechanical system will pass through every mechanically possible state. This assumption was called the ergodic hypothesis. In the beginning of the century, doubts were expressed whether this idea could be given a self-consistent formulation that would allow it to be applicable to actual mechanical systems with an arbitrarily large but finite number

---

[19] See Uhlenbeck and Ford, *Lectures on Statistical Mechanics*, American Mathematical Society (Providence, RI, 1963). Boltzmann's solution to the puzzle was the following: Paths relative to which the entropy *always* increases are relatively rare. In most cases, the entropy will fluctuate. Therefore, the content of the second law is that even when we encounter a path whose entropy decreases in some time segment, it is likely that in both earlier and later time segments the entropy on this path was and will be higher. This formulation is consistent with Loschmidt's observation.

of particles. The problem, as Plancharel and Rosenthal pointed out,[20] was that a one-dimensional trajectory cannot fill a space of higher dimensionality.

Finally, a need was felt to explain the astounding empirical success of Gibbs's methods. This puzzle is particularly annoying to those who criticize Gibbs's pragmatic methodology. Another related problem concerns the relations between Gibbs's theory and similar pragmatic themes that appear in the context of the justification of quantum mechanics.

Now we can begin with a review of the ergodic approach. We shall first introduce some mathematical concepts and theorems from a mathematical theory called ergodic theory. Next we shall show how these concepts and theorems are relevant to physics and formulate the ergodic approach to SM. Finally, we shall see whether the ergodic approach can yield solutions to any or all of the problems that we have mentioned. As we shall find out, the ergodic approach, as it is usually defined, cannot give satisfactory answers to every question. Therefore, we shall have to discuss the fruitfulness of various possible extensions of the ergodic approach.

2.3. AN INTRODUCTION TO ERGODIC THEORY

We begin with a brief summary of the basic concepts of the ergodic theory. One should remember that this theory is a mathematical construct that may be studied independently of its applications to SM.[21] However, because the ergodic theory was developed as an abstraction from the mathematical representation of isolated many-particle systems, we may follow the process of abstraction and so gain an insight into the concepts and definitions that we shall encounter later on.

---

[20] Rosenthal and Plancherel proved the impossibility of the "simple" ergodic hypothesis their "Beweis der Unmoglichkeit ergodischer Gassysteme," *Ann. der Physik* 4(42) (1913), pp. 796–806 and "Proof of the Impossibility of Ergodic Mechanical Systems," *Ann. der Phys.* 4(24) (1913), pp. 1061–1063, respectively. It is less clear whether anyone has ever defended this hypothesis explicitly. (Rosenthal mentions the review of the Ehrenfests as the source of the formulation of EH, and Plancherel does not mention the source of the formulation at all.) For a good article on the subject, see Brush, "Proof of the Impossibility of Ergodic Systems: The 1913 Papers of Rosenthal and Plancherel," *Transport Theory Statist. Phys.* I (1971). We shall continue the discussion on some of these issues in Chapter 3.

[21] For a survey of ergodic theory, see Walters, *An Introduction to Ergodic Theory* Springer–Verlag (New York, 1982).

A state of an n-particle system M depends on $6n$ parameters. Each of the particles must be assigned three position and three momentum coordinates. When these parameters $p_1, \ldots p_{3n}, q_1, \ldots q_{3n}$ are assigned, the state of the system is fixed. Therefore, it is possible to represent each state as a point in a $6n$-dimensional space that is isomorphic to a subspace of the Euclidean E^{6n}. This space is known as the *phase space* of M.

One of the standard ways of describing the evolution of M is by stating its Hamiltonian $H(p_1, \ldots, p_{3n}, q_1, \ldots, q_{3n})$ using the equations

$$\partial H / \partial p_i = dq_i / dt, \qquad \partial H / \partial q_i = -dp_i / dt.$$

The Hamiltonian can be described as a transformation from E^{6n} to itself, which determines for every state its dynamic evolution after an infinitesimal time interval. This means that every state lies on a unique trajectory that is determined by its Hamiltonian. Because the solutions are required to be reversible in time,[22] every state has a unique past trajectory as well. Therefore, a Hamiltonian can be thought of as a decomposition of the phase space into a system of mutually exclusive and exhaustive trajectories that extend to the infinite future from the infinite past. Every possible state of M lies on exactly one such trajectory. The exclusiveness is the result of the uniqueness of the solutions of H, and the exhaustiveness is the result of the existence of solutions for H.

One of the most important results concerning Hamiltonian systems is Liouville's theorem. Let A_t be a set of states, and let $A_{t+\Delta t}$ be the set of the future evolutions of each and every one of the members of A_t after Δt. Liouville's theorem states that the volumes of A_t and $A_{t+\Delta t}$ are the same.[23] This property of Hamiltonian systems is known as *stationarity* or *incompressibility*.

The first step in the mathematical abstraction from the theory of Hamiltonian systems is to abstract from the Hamiltonian only its essential features, namely, the trajectory that it assigns to every state. This abstraction is effected in two stages. First, we look at the Hamiltonian as a topological group of operators $\{U_t\}_{t \in R}$; in other words, for every $t \in R$ (R is the set of real numbers), $U_t : E^{6n} \to E^{6n}$ assigns

---

[22] The backward motion of a mechanical system is a possible mechanical motion as well.

[23] We assume that both A_t and $A_{t+\Delta t}$ are measurable.

to every state $s \in E$ its future at t. The composition $U_t U_{t'}$ is defined as $U_{t+t'}$. The unit element U_0 is simply the "present time operator," that is, for every $s \in E^{6n}$, $U_0(s) = s$. It is easy to see that $U_t U_{-t} = U_0$. Our next step is to replace the continuous structure of $\{U_t\}_{t \in R}$ with a discrete $\{U_t\}_{t \in Z}$ (Z is the set of integers). We may do so by defining a transformation $T: E^{6n} \to E^{6n}$ such that, for all $s \in E$, $T(s)$ is the future of s after a short interval of time; we then identify $\{U_t\}_{t \in Z}$ with the group of iterations T^i.

Let $s \in E^{6n}$ be an arbitrary state. The trajectory of s under T is $O_T(s) = \{s, T(s), T^2(s), \ldots\}$. Let A be an arbitrary set of states. The time average that M stays in A given its present state s is $O_T(s, A) = \lim_{n \to \infty} \frac{1}{n} \Sigma_{i=1} X_A(T^i(s))$ [X_A is the characteristic function of A; that is, $X_A(s) = 1$ if $s \in A$, and $X_A(s) = 0$ otherwise.] Observe that $O_T(s, A)$ can be thought of as the frequency in which $O_T(s)$ assumes states that are located in A.

To understand the significance of the concept of time averages for SM, let us look at one important example. Let A_E be the set of states that are compatible with macroscopic equilibrium.[24] Recall that: When M is made of many particles, the volume of A_E is overwhelmingly large.[25] We may now formulate the following questions:

1. Given $s \in S$, what is $O_T(s, A_E)$? Or, given that M is in the state s now, how often will M assume states that are compatible with macroscopic equilibrium?
2. Under which circumstances will $O_T(s, A_E)$ be close to 1?
3. Under which circumstances will $O_T(s, A_E)$ be independent of the choice of s?

Let us assume that $O_T(s, A_E)$ is large, independently of s, in a particular case of interest. In this case we are well on our way to justifying our expectations that M will reach a persisting equilibrium independent of its initial conditions. As we shall see shortly, ergodic systems are those for which the above-mentioned conditions obtain assumption. This explains why the search for proofs of ergodicity is the

[24] It is impossible to determine, just by looking at the location of a state in E^{6n}, whether or not it is an equilibrium state. However, if we look at the representation of s as n points in the six-dimensional space we discussed in the last chapter, we shall be able to determine whether it belongs to A_E or not.
[25] See Chapter 1.

75

most important foundational task from the ergodic theoretic point of view.

So far we have abstracted from the properties of the Hamiltonian and we have arrived at the transformation T. Ergodic theory proper proceeds with a further abstraction process that also involves the phase space. In ergodic theory, the basic mathematical structure is a quadruple $\langle S, B, m, T \rangle$, which is called a *dynamic system*. S is a set of possible states of M, B is a sigma algebra of subsets of S,[26] $m : B \rightarrow R$ is a measure,[27] and $T : S \rightarrow S$ is the evolution transformation we are familiar with.

Remarks

1. *We assume that* T *is 1-1 but not necessarily onto. That is why we shall make use of the inverse* T^{-1} *in the definitions instead of* T. *When* T *is onto, there is no need to do so.*
2. S *is the set of states possible for* M, *in an objective sense of possibility.*
3. B *is the "descriptive apparatus" of* M.[28] *It is obtained as a generalization of the Lebesgue measurable sets of* E^{6n}.
4. T *is supposed to be a measurable transformation. That is, for all* $A \subset S$, $A \in B$ *if and only if (iff)* $T^1(A) \in B$. *As we remarked earlier,* T *is an abstraction from the Hamiltonian of* M.
5. *The measure* m *is an abstraction of the volume measure on* E^{6n}. *If there is a topological structure on* S *as well, it is common to require that* m *be regular with respect to this structure, that is,* $m(A) = inf\{m(C) \,|\, C \supseteq A$ *and* C *is open}.*
6. *Most textbooks on ergodic theory require that* m *be* T-*invariant; that is, for all* $A \in B$, $m(A) = m(T^{-1}(A))$. *This property generalizes the property of stationarity, which can be proven for Hamiltonian systems using Liouville's theorem. This requirement is objectionable from a subjectivist point of view. Some subjectivists also object to*

---

[26] A σ-algebra is a set of subsets of S that is closed under countable intersections and complementation.
[27] A measure is an additive set function such that, if $A \cap B = \emptyset$, then $m(A \cup B) = m(A) + m(B)$. A σ-additive measure has the additional feature that $m(\bigcup_{i=1}^{n} A_n) = \sum_{i=1}^{n} m(A_n)$, where $\{A_n\}$ is a denumerable collection of mutually exclusive sets.
[28] Different choices of B may yield nontrivially different theories. See Chapter 1.

the requirement that m *should be sigma-additive; instead, they merely require finite additivity.*[29]

7. *We shall assume often that* M *is a closed system that is contained in a finite volume. More precisely, we will assume that* S *is a bounded, connected subset of* E^{6n}.[30] *It is common to generalize this property by requiring that* m(S) < ∞.[31] *In fact, we shall require that* m(S) = 1, *whereby* m *becomes a sigma-additive,* T-*invariant, measurable probability function.*

8. *We shall refer to* ⟨S,B,m,T⟩ *as a measure-preserving system.*

9. *When a property holds for a subset of measure one, we shall say that it holds* m-*almost always, where* m *is the relevant measure.*

10. *Two measures,* m *and* m′, *are considered equivalent iff they assign measure zero to the same collection of sets. (We write* m ∼ m′.) *If the collection* A^0_m = {A ∈ B: m(A) = 0} *is included in* $A^0_{m'}$ = {A ∈ B: m′(A) = 0}, *we say that* m *is absolutely continuous with respect to* m′. *(We write* m < m′.) *The Radon–Nikodym theorem states that when* m < m′, *there is a positive function* f *that is* m-*measurable and* m′ = ∫fdm.

Definition. T *is ergodic iff, for all* A ∈ B, $T^{-1}(A)$ = A *implies that* m(A) = 0 *or* m(A) = 1. *(Or, equivalently,* m *is an ergodic measure iff the above condition holds.)*

Theorem I (Birkhoff 1931). *If* T *is ergodic, then for all* f ∈ L^2, *and for* m *almost every* s ∈ S,

$$lim_{n \to \infty} \frac{1}{n} \sum_{i=1}^{n} f(T^i(s)) = f*,$$

where f* ∈ L^2. *(L^2 is the space of the square-integrable functions.*[32])

Corollary I.[33] *If* T *is ergodic, for every* A ∈ B *and for* m-*almost every* s ∈ S,

---

[29] For a discussion of these issues, see de Finetti, *Theory of Probability*, vol. I, Wiley (New York, 1970), pp. 228–235.

[30] More precisely, the abstract requirement is that there are no wandering sets of positive measure, that is, no such sets A such that, for all n, A ∩ $T^n(A)$ = 0.

[31] In topological spaces "finite" sets are compact. Such sets can be "covered" by a finite number of open neighborhoods.

[32] A function f is a square-integrable function if the integral of f^2 exists and it is finite.

[33] For a proof, see Walters, ibid, p. 41.

$$lim_{n\to\infty} \frac{1}{n} \sum_{i=1}^{n} X_A(T^i(s)) = m(A).$$

In plain words, if T is ergodic, the time averages $O_T(s,A)$ are the same as the measure of A for every $A \in B$ and for almost every $s \in S$. In particular, if $m(A_E) = 1$ and T is ergodic, $O_T(s,A_E) = 1$ almost always.

Theorem II. *If* T *is ergodic, for every* A,C $\in$ B,

$$lim_{n\to\infty} \frac{1}{n} \sum_{i=1}^{n} m(T^{-i}(C) \cap A) = m(A)m(C).$$

Corollary II

$$\frac{1}{n} \sum_{i=1}^{n} X_A(T^i(S))X_C \to m(A)X_C, \qquad \textit{for almost every } s \in S.$$

The significance of Theorem II and its corollary to SM can be gathered from the following considerations. The theorem states that the ergodicity of T implies that, when we calculate $O_T(s,C)$ using only s that are members of A, the result will not be biased, on the average, because the C are distributed among the A in the same proportion that they are distributed among the general population. This is precisely why we can discard additional information (e.g., that $s \in A$) when we predict the future state of an ergodic system. In particular, this explains why any extra knowledge concerning M will not affect our predictions concerning whether and how often M will assume an equilibrium state.[34] Note that ergodicity is a nontrivial condition on the average dynamics of M. Therefore, the presence of ergodicity constitutes a nontrivial explanation for what is considered an anomaly from a standard Bayesian point of view, namely, for the fact that available information can always be presumed irrelevant for predicting equilibrium states.

As we remarked earlier, the concept of a dynamic system was obtained by Poincaré and Birkhoff as a generalization or abstraction from Hamiltonian systems. However, when we study dynamic systems as abstract mathematical objects, we discover that there are systems of

---

[34] A stronger condition of asymptotic irrelevance is called mixing. See Section 2.4 in this chapter.

different origins that can all be represented formally as dynamic systems. One example that is very important to our concerns is the construction of a dynamic system whose origin is a stochastic process. Let us follow this construction with some detail.

The better known example of a stochastic process is a coin that is tossed repeatedly. Such a process can be described as a dynamic system in the following way:

1. As the space of states S we shall take all of the (doubly) infinite sequences of heads and tails. Each sequence $\ldots w_{-1}, w_0, w_1, \ldots$ is an exhaustive description of one of the possible outcomes of the infinite coin-tossing process.

2. As our descriptive apparatus B, we shall take the σ-algebra generated by the cylindrical sets. As you may recall, cylindrical sets define sets of sequences by stipulating that their ith coordinate should have a particular value. For example, "the set of all sequences that have heads in their 137th place" is a cylindrical set.

3. As T we may take the shift transformation. Recall that the shift is defined as $T(\ldots w_{-1}, w_0, w_1, \ldots) = \ldots w_0, w_1, w_2, \ldots$. The shift assigns to every sequence a different sequence that has the same elements ordered in the same way but parametrized differently. T can be thought of as the "unfolding in time" of a particular sequence. The shift reveals the identity of the result of the next toss at each moment.

4. Let $A \in B$. For example, take

$$A_{137} = \{\ldots w_{-1}, w_0, w_1, \ldots \mid w_{137} = \text{heads}\}.$$

A_{137}, of course, is a cylindrical set. The quantity $m(A_{137})$ measures the "relative proportion" of sequences that have heads in the 137th coordinate. We say that m is stationary if it is shift-invariant, that is, if m assigns the same probability to the event that heads appears in the 137th coordinate as it assigns to the appearance of heads in any other coordinate. It is fairly common to read in textbooks on probability that coin-tossing processes have stationary measures if the mechanism that results remains the same. In such cases, they say, there is a fixed probability that any toss will yield heads independently of the precise moment of its execution. This rationale for stationarity is unacceptable for subjectivists. For them, the fact that the tossing mechanism remains the same is besides the point. We shall return to this point later.

5. The conditional measure $m(A/A')$ is interpreted, as always, as the "relative proportion" of the sequences that have the property A among those that have the property A'. For example, $m(A_{137}/A_{246})$ is the relative proportion of the sequences that have heads in the 137th place among those that have heads in the 246th place. Measures for which for all $A, A' \in B$, $m(A/A') = m(A)$ are called Bernoulli measures.

6. For the definition of an arbitrary stochastic process, replace the set {heads, tails} with a fixed partition $p_1, \ldots p_n$. The assumption is that at each moment exactly one of the p_i will happen. The rest of the generalization is rather straightforward.

7. The trajectory $O_T(s) = \ldots T^{-1}(s), s, T(s), \ldots$ of an element of a stochastic process that is shifted is the infinite sequence of infinite sequences $(\ldots w_{-1}, w_0, w_1, \ldots)$, $(\ldots w_0, w_1, w_2, \ldots), \ldots,$ $(\ldots w_n, w_{n+1}, w_{n+2}, \ldots), \ldots$. The time average $O_T(s, A_{137})$ is the measure of the relative frequency of the sequences in $O_T(s)$ that have heads in the 137th coordinate. Note that $O_T(s, A_{137}) = O_T(T(s), A_{136})$ because $O_T(s)$ has the same set of sequences but whose coordinates were shifted to the left once.

8. Ergodicity in the shift space means that the "future" coordinates of a sequence are asymptotically independent of the present coordinates. (The notion of asymptotic independence and its relation to ergodicity were discussed in Theorem II and its corollary.) Therefore, Bernoulli measures are clearly ergodic, because independence implies asymptotic independence.

2.4. FREQUENTISM AND THE ERGODIC APPROACH

It is often stated that the ergodic approach takes its motivation from a frequentist conception of probability in the context of SM. Now we can substantiate the claim, at least in part.

From a frequentist perspective, the probability of an event is defined as the relative frequency with which the event happens in a sequence of random independent trials. The frequency is taken to be an objective property of the sequence of trials. Therefore, many writers think of frequentism as an objectivist conception of probability. For example, the frequentists define the probability of obtaining heads on a toss of a coin as the quantity

$$P(H) = \lim_{n \to \infty} \frac{1}{n} \sum_{i=1}^{n} X_H(w_i),$$

where w_i is the result of the ith toss in a sequence of "independent random trials" $w_0, w_1, \ldots$ and $X_H(w_i) = 1$ if $w_i = H$, $X_H(w_i) = 0$ otherwise. This definition makes $P(H)$ an objective property of the w_i.

This definition is quite reminiscent of the definition of time averages. In fact, if we could regard the sequence $O_T(s) = \{s, T(s), T^2(s), \ldots\}$ for an arbitrary $s \in S$ to be a sequence of random independent trials, time averages could yield a concept of probability that would be very much within the spirit of the frequentist conception. Observe, however, that the process that generates the $T^i(s)$ might be a completely deterministic process. Therefore, if the notion of independent trials is taken to imply that the result of w_i is causally independent of the result of w_{i-1}, the sequence $O_T(s)$ is clearly *not* necessarily a sequence of independent trials. If, on the other hand, the relevant notion of independence is probabilistic, we need a prior conception of probability to decide whether or not $O_T(s)$ is a sequence of independent trials. (Note that, without the requirement that the definition of probability be given relative to a sequence of independent trials, there is no reason to expect that the frequencies in two different random sequences will be the same.)[35] The ergodic conception of time averages meets the second horn of the dilemma; that is, it forfeits the claim that probabilities may be *defined* as relative frequencies. Consequently, there is no attempt to define a random sequence. Instead, there is a formulation of a condition that guarantees that the frequencies in different sequences will be almost always the same. This condition is, of course, ergodicity. As we have seen in Corollary I, if T is ergodic, for every $A \in B$ the time averages $O_T(s, A)$ are almost always the same. Notice that the presence of m in the definition of ergodicity prohibits us from using $O_T(s, A)$ as the *definition* of $m(A)$. However, when we can prove that a system $\langle S, B, T, m \rangle$ is ergodic, we can use $O_T(s, A)$ as an estimate of $m(A)$, and vice versa. This points to the important role of frequencies within the ergodic approach.

At this point the informed reader may raise an eyebrow and ask the following question. The concept of a random sequence was defined by von Mises and his followers using the property of invariance with

---

[35] The requirement that the frequencies be the same in different random sequences of trials of the same kind amounts to requiring that it must be possible to determine them by a repeatable experiment. The requirement is especially important when the concept of relative frequencies is interpreted as the operational definition of probability.

respect to place selection.[36] Is there any relation between ergodicity and this property? The short answer to the question is NO. Ergodicity does not guarantee that when we select a subsequence $\{T^{i_1}(s),$ $T^{i_2}(s), \ldots\}$, where the i_j form an infinite increasing set of indexes, we shall have

$$\lim_{n \to \infty} \frac{1}{n} \sum_{j=1}^{n} X_A(T^{i_j}(s)) = \lim_{n \to \infty} \frac{1}{n} \sum_{i=1}^{n} X_A(T^i(s)). \quad (*)$$

In that sense, the connections between von Mises's theory and the ergodic approach are slightly weaker than they first appear.[37] However, what does guarantee the property (*) is a condition called *mixing*, which is stronger than ergodicity but is still related to it.

Definition. T *is mixing iff*

$$lim_{n \to \infty} m(T^{-n}(A) \cap C) = m(A)m(C).$$

This stronger sense of convergence guarantees that there will be no infinite subsequences of i_j that will converge to a separate limit. Because there are many cases where ergodic transformations are also mixing,[38] the ergodic approach does have a genuine connection with frequentism.[39]

---

[36] For the exposition of von Mises's original argument, see his *Probability, Statistics and Truth*, Allen and Unwin (London, 1928). Other frequentists, such as Reichenbach, did not rely on the concept of place selection. A more sophisticated characterization of random selection was developed by Martin-Lof, "The Definition of Random Sequences," *Inform. Control* 9 (1966).

[37] See Von Plato, "The Significance of the Ergodic Decomposition Theorem of Stationary Measures for the Interpretation of Probability," *Synthese* 53 (1983), pp. 429–433: "The ergodic decomposition of stationary measures . . . clarifies the idea of frequentist probability. Since ergodicity is equivalent to the condition that sequences of events have identical asymptotic properties, especially limits of relative frequencies, it corresponds to a measure-theoretic notion of frequentist probability."

[38] See Walters, p. 50, for such an example. For a discussion on the difference between ergodicity and mixing, see Wightman, "Statistical Mechanics and Ergodic Theory," in *Statistical Mechanics at the Turn of the Decade*, ed. Cohen, Dekker (New York, 1970), pp. 11–16.

[39] For some purposes, even weak mixing will suffice. For a definition of weak mixing, see Walters, p. 40.

2.5. APPLICATION OF THE ERGODIC THEORY TO THE THEORY OF HAMILTONIAN SYSTEMS

Having become acquainted with the mathematical apparatus of the ergodic theory, we can now try to see how the various definitions and theorems that we presented apply to the physical theory of Hamiltonian systems. Of particular importance are the implications of the ergodic theorem and its corollaries to SM. The discussion will be centered on two points that need to be clarified for us to understand the significance of the ergodic theorem to physics. The first is the idea that B, the descriptive apparatus, should be chosen so as to represent a "coarse grain" description of all of the possible results of the measurement of a macroscopic parameter. The second idea is that we should think of $O_T(s,A)$ as a particular result of a measurement process that requires an appreciable time to perform.

Let us begin with a discussion on the choice of a descriptive apparatus B. A clarification of this point is not a trivial matter, because there are at least two different choices that can be made concerning B, and each choice implies a different interpretation of the ergodic theorem.

The first choice of B is inspired by Gibbs's writings. We think of B as a collection of infinitesimal cells in the phase space. Relative to this choice we obtain the "fine grain" ergodic theorem, a theorem that is not devoid of interest to the physicist. Ergodicity implies, for example, that arbitrarily small areas in the phase space will be traversed by almost every trajectory. This fact is already sufficient for the proof of Poincaré's recurrence theorem. If almost every trajectory passes through *every* small area, it must be true, in particular, that almost every trajectory passes through small neighborhoods of its origin. In fact, the ergodic theorem implies that almost every trajectory will pass through neighborhoods of its origin infinitely many times. Moreover, the theorem implies that the average frequency in which the trajectories will pass through an area is proportional to the size of the area.

The second possibility is to think of B as a collection of the possible results of a macroscopic measurement, each of which is consistent with a large number of different microstates. This idea is an important one. The coarseness of the division of the phase space corresponds to the limitations of our discriminatory capacity or to the theoretical and practical constraints that prohibit us from registering empirical states with an arbitrarily high degree of accuracy. This interpretation of B therefore incorporates the insights of the theory of macroscopic measurements. Such theory is of particular importance for the discussion on quantum phenomena becaus the uncertainty relations of quantum mechanics can be introduced as limitations on the size of the phase cells.[40] But it is important to realize that the fact that we are not infinitely discriminating creatures has important consequences even in classical mechanics. There are physicists who think that this fact alone is sufficient for explaining irreversibility.[41]

Suppose that we can justify the idea that measurement processes should be represented as time averages. (We shall discuss the status of this assumption shortly.) Such a conception will motivate the following "coarse grain" form of the ergodic theorem. When we perform macroscopic measurements on an ergodic system, the frequency in which particular results will be realized is, almost certainly, the same as the microcanonical measure of the set of microstates that are compatible with the experimental result. In other words, the "coarse grain" ergodic theorem allows us to think of the results of the measurement as time averages taken over a long period of time while at the same time permitting us to use the phase averages for the purpose of prediction. This idea has a clear foundational significance.

Why, though, should we think of the results of measurement as time averages? The most familiar argument is the following. The time that it takes to make a measurement, the argument goes, is considerably longer than the time scale of microscopic processes. Therefore, the process of measurement cannot be conceptualized as one that allows us to register the precise physical state of the system at a given time. Rather, we should think of the measurement process as a long inter-

---

[40] In fact, the procedure of coarse graining is absolutely necessary in quantum SM for other reasons as well. The fine-grain quantum ergodic theorem raises more objections than its classical counterpart. See Farquhar, pp. 142–168. See also footnote 48.

[41] This was Jaynes's view; see Chapter 1 of this book. Other writers tried to import some features of this coarseness into the discussion on fine-grain dynamics. Krylov, for example, claimed that our initial ensembles can be constructed only of "simple" sets that become "complicated" after the passage of time. In this manner he attempted to account for the asymmetry between the present and the future. See Krylov, *Works on the Foundations of Statistical Mechanics*, Princeton Univ. Press, (New Jersey, 1979). For a presentation of Krylov's ideas and discussion on their philosophical significance, see Sklar, *Physics and Chance*, pp. 262–269. We shall remark on the notion of "simple sets" in Chapter 4.

action with the system during which time the system continually evolves. What we register is the impact of the system on the measuring device during the whole period of interaction. Now, as the system evolves, the parameter we measure may assume different values, and some of these values may appear more frequently than others. Those values that appear more frequently (all other things being equal) will leave a more distinct mark on the measuring device than the "very rare" values. Therefore what we are describing in fact is a natural process that resembles the operation of time averaging. It is a process that registers the relative weights of the various sets of time points in which the various values are instantiated. When the difference between the microscopic and the measurement time scales is very large, the argument concludes, we are permitted to think of the measurement process as an infinite time average. Where ergodic systems are concerned, the time averages will almost always be the same. This property, then, allows us to think of measurements as repeatable processes.[42]

There are some standard objections to the identification of measurement processes with infinite time averages.

1. In some cases, the difference between the microscopic time scale and the scale of measurement is not very great. In particular, when we wish to take the effect of the interaction between the molecules and the walls of the container into account, we have to consider the time that it takes a molecule to reach the walls of the container. That of course makes the microscopic time scale much longer.

2. If single measurements are conceptualized as infinite time averages, how shall we represent macroscopically observable fluctuations? For example, how can we model phenomena such as the Brownian motion? A failure to address this point will be particularly troubling because, as you may recall, the Brownian motion was historically one of the most convincing reasons for accepting the framework of SM.

3. More generally, ergodicity may be too stringent a condition. After all, every now and then we do discover that some measurement

---

[42] We are assuming, for the time being, that the interaction with the measuring device is "very weak"; otherwise, we cannot apply the ergodic theorem the way we have stated it so far. Later in the chapter we shall discuss the meaning of this assumption.

procedure is biased. In these cases, we would like to intervene with the measurement process and "correct" it. But, if we base our theory of measurement on the assumption that "most" macroscopic processes are ergodic, interventions (or even calibrations) do not seem to be possible.

A partial solution to these objections was obtained by Hopf from the following considerations.[43] Hopf suggested that we continue to think of a single act of measurement as an infinite time-averaging process. However, he suggested that when we ask for the rationale for replacing space–time averages with time averages, we should think of taking the average of an infinite sequence of single measurements that happen one after the other. In other words, if the result of a single A-measurement is

$$\lim_{n \to \infty} \frac{1}{n} \sum_{i=1}^{n} X_A(T^i(s)) = f_t(A,s) = f_t,$$

then the infinite average of the single results is

$$\lim_{t \to \infty} \frac{1}{t} \sum_{i=1}^{t} f_t.$$

Observe that this picture introduces some new difficulties.

1. Hopf's idea may apply only when the scale difference between the macro- and the microprocesses is truly very large.
2. When we average the results of different measurements, we have to require $\lim_{t \to \infty} \frac{1}{t} \sum_{i=1}^{n} (m(A) - f_t)^2$, or the dispersion, to be small. This property is not a consequence of ergodicity; it is a stronger condition that may prove to be even more difficult to ascertain in particular cases of interest.
3. Measuring devices that continue to perform measurements on a system during an indefinite period of time are not very easy to come by. It is not clear how to apply Hopf's idea to more common situations.

---

[43] See Hopf, *Ergodentheorie*, Chelsea (New York, 1937) and "On Causality, Statistics and Probability," *J. Math Phys.* 17 (1934), pp. 51–102.

2.6. HAS THE ERGODIC APPROACH SOLVED THE FOUNDATIONAL PROBLEMS OF SM?

Now it is time to go back to the discussion on the fundamental problems of SM. Now that we have an understanding of the relevance of the ergodic theorem to the study of Hamiltonian systems, we can attempt to find out whether the ideas we have developed so far can help us to solve these problems. Our conclusion, as we shall see, is a mixed bag of negative and positive assessments.

Let us begin with a group of remarks concerning the issue of the mathematical formulation of the ergodic hypothesis and its physical realizability.

1. The replacement of the original ergodic hypothesis with the quasi-ergodic hypothesis yielded a mathematically consistent formulation of the ideas of Maxwell, Boltzmann, Gibbs, and Einstein. While the ergodic hypothesis can be true only of the most trivial systems, the quasi-ergodic hypothesis, on the other hand, can be satisfied by systems with any number of degrees of freedom.[44]

2. The next important question is whether actual Hamiltonian systems are in fact ergodic. Ulam and Oxtoby proved that ergodic systems are dense among continuous homeomorphisms.[45] This means that arbitrary continuous transformations can be approximated by ergodic transformations. For a while, this was taken as an indication that ergodicity is the rule rather than an exception. Gamow, for example, wrote that the result proved "that in a certain sense almost every continuous transformation is metrically transitive."[46] This conclusion, however, was widely rejected after Kolmogorov, Arnold, and Moser (KAM) proved that, among the differential homeomorphisms, which are subsets of the continuous homeomorphisms with more structural features, there was a significant set of nonergodic transformations.[47] Such transformations describe the motions of stable systems that have quasi-periodic trajectories. When systems

---

[44] We shall continue the discussion on this issue in Chapter 3.
[45] See Ulam and Oxtoby, *Ann. Math.* 42 (1942).
[46] See his translator's note in Khinchin's book, ibid, p. 54.
[47] For a review of the KAM theory, see Walker and Ford, "Amplitude Instability and Ergodic Behavior for Conservative Non-Linear Oscillator Systems," *Phy. Rev. V* 189(1) (1969). See also the discussion in Sklar, *Physics and Chance*, pp. 169–175.

with a large number of degrees of freedom are concerned, one should not expect quasi-periodic behavior. In this sense, the KAM theorem does not contradict the ergodic approach. Nevertheless, it is now clearer that arguments that establish the ergodicity of Hamiltonian systems of the type that we encounter in SM have to be more refined. After all, Hamiltonian systems with a small number of degrees of freedom (the planetary system, for example) are not expected to be ergodic.

3. More generally, one of the outcomes of the research on dynamic systems was that the notion of instability became more central to the discussion on the reasons for stochastic behavior. Unstable systems are characterized as those whose trajectories diverge rapidly. There are interesting connections between ergodicity and instability, and we shall return to the discussion on this topic in Chapter 4.

Let us now address a point that was important to many of the originators of the ergodic approach, namely, whether defining probabilities in terms of time averages allows us to think of probabilities as physical observables. Again, the verdict is not clear.

1. On the one hand, the time average $O_T(s, A)$ does seem to have a physical meaning: It is the relative frequency in which the trajectory emanating from s stays in the set A.

2. Many writers, however, objected to the identification of time averages with the long-run behavior. Strictly speaking, time averages are taken over an infinitely long period. In most cases, we do not have proofs that the finite time averages will converge to the limit rapidly.

3. More important, though, is the fact that we may expect different time averages relative to different trajectories. To be sure, ergodicity guarantees that almost every trajectory will have the same time averages. Therefore, one may feel justified to identify probabilities with the "dominant" time averages, but this suggestion leads to a circular argument: We started by attempting to define probabilities using time averages and ended up relying on a previous definition of probability to define time averages.

4. The ergodic theorist may attempt to respond to the last argument by pointing to the fact that, if a system is ergodic relative to probability function m and m' is another ergodic probability function that is not singular with respect to m, then the dominant time averages

relative to m are the same as the dominant time averages relative to m'. (Two ergodic measures are either singular or equivalent.) Therefore, as long as there is agreement over either (a) whether a given system has nontrivial invariants of the motion or not (that is, whether it is decomposable or not) or (b) whether a set of states has measure zero or not, there will be an agreement concerning the notion of the dominant time averages.

5. The opponent of the ergodic approach is not likely to be impressed with the above-mentioned retort. First, we need a prior probabilistic notion to define the concept of an invariant of the motion. Second, even if there is widespread agreement over the choice of sets with probability zero, such agreement might be simply a deeply rooted convention. Do we want the definitions of our scientific concepts to depend on such conventions?

Next, let us address a question that seems to be very central to the whole discussion: Do we have a good argument for replacing the laborious time averages with the easily calculable space averages? Here the positive contributions of the ergodic approach are more clearly visible.

1. Recall that ergodic systems have the following property. For every $A \in B$ and for m-almost every $s \in S$,

$$\lim_{n \to \infty} \frac{1}{n} \sum_{i=1}^{n} X_A(T^i(s)) = m(A).$$

This property expresses the sense that ergodicity guarantees us an equality between time averages and space averages.

2. One should note, however, that ergodicity is a special property that holds only in special cases. Therefore, as long as we do not have a *proof* that nonergodic systems are not encountered in practice, the ergodic approach is based on an unsubstantiated assumption. In addition, even where ergodic systems are concerned, the equality between time averages and phase averages holds "up to a set of measure zero." As we remarked already, this expression cannot be given an operational interpretation without presupposing prior probabilistic concepts.[48]

---

[48] In quantum SM, a general equality of the time averages with the phase averages, regardless of the initial state, implies the nondegeneracy of the energy eigenvalues.

3. Khinchin introduced another rationale for the replacement of time averages with space averages. His argument was the following. In physics we encounter dynamic systems that have various special properties. First, the systems of relevance have a large number of degrees of freedom. Second, the parameters of interest are representable as sum functions. (A function that is a sum of functions that are dependent only on the dynamic variables of a single molecule are called, by Khinchin, sum functions.) Therefore, when dealing with the foundations of SM, we may justify the replacement of time averages with space averages only with regard to such systems. Using these special properties, Khinchin proved that the dispersion of the sum functions from the mean must be small. More generally, he showed how to apply the law of large numbers to the case of systems with many degrees of freedom.[49] An important feature of Khinchin's treatment is that it demonstrates that the phase space of a many-particle system may include a set of trajectories with a small but positive measure, all the members of which are "deviant trajectories" with time averages that are different from the phase average.

4. The idea of reintroducing into the ergodic approach the fact that in SM we are dealing with systems with many degrees of freedom may seem objectionable to some people, especially those for whom the number of degrees of freedom, taken by itself, does not have any independent explanatory power. Furthermore, as we shall see shortly, Khinchin's argument presupposes substantive statistical assumptions.

Our next topic is the Stosszahlansatz. More precisely, we shall try to find out whether the ergodic approach sheds light on the issue of the consistency of the assumption that the collisions between the molecules annihilate statistical correlations and the underlying deterministic mechanical framework.

This condition furnishes us with a fairly good quantum analog to metric indecomposability. However, the condition of the nondegeneracy of the energy eigenvalues refers to all of the energy values while metric indecomposability refers to fixed energy levels. Moreover, the time averages continue to refer to the weights of the initial state. This additional dependence prevents the time averages from being identical with the microcanonical averages. (See Jancel, ibid, pp. 62–65 and Farquhar, ibid, pp. 142–146.)

[49] See Khinchin, *Mathematical Foundations of Statistical Mechanics*, Dover (New York, 1949), pp. 62–69.

1. As we remarked, ergodic systems have the following property: For every $A, C \in B$,

$$\lim_{n \to \infty} \frac{1}{n} \sum_{i=1}^{n} m(T^{-i}(C) \cap A) = m(A)m(C).$$

Therefore, we may say informally that when ergodic systems are concerned, on the average, any correlation between A and C will be canceled in the "infinite long run." As we remarked earlier, systems of SM origin often have an even stronger property called mixing. For such systems,

$$\lim_{n \to \infty} m(T^{-n}(C) \cap A) = m(A)m(C)$$

for every A and C in B.

3. Khinchin offered an interesting argument that suggests a connection between ergodicity and the Stosszahlansatz. Again, his argument depends on the special properties of systems of SM origin. First, Khinchin characterizes molecular chaos as the decay of the coefficient of the correlation between T^n and T^{n+i} as they apply to individual molecules.[50] In other words, molecular chaos obtains when the knowledge of the precise state of a single molecule at n will not be a good predictor of the state of the same molecule at $n + i$. Next, Khinchin argues that when systems with many molecules are concerned there is indeed a decay of the correlation between T^n and T^{n+i} as they apply to individual particles. The collisions between particles will render knowledge of, for example, the energy of a single particle at t useless for the purpose of predicting the energy of the particle at a future point $t + \Delta t$. Now, the decay of correlation implies ergodicity, and Khinchin proves a theorem to this effect.[51] So the motion of the individual particles is ergodic. Remember, though, that in the case of SM, we always encounter sum functions. For such functions, the ergodicity of the sum function follows from the ergodicity of the individual particles.

The next set of remarks concerns Poincaré's recurrence theorem.

---

[50] See Khinchin, *Mathematical Foundations*, p. 67.
[51] Ibid, p. 68.

1. It is clear that ergodicity is compatible with "almost sure recurrence." In fact, as we remarked earlier, ergodicity implies recurrence. This point was emphasized by Birkhoff in his original proof of the ergodic theorem.[52]

2. We also know that the recurrence time of the universe as a whole is considerably longer than the life span of the universe. In this sense, it does not present an immediate contradiction to the belief in the general trend toward disorder.

Our next topic is the issue of irreversibility. Here the contribution of the ergodic approach is most debatable. Even some of the proponents of the ergodic approach admit that a separate account of the irreversible character of macroscopic systems may be necessary.[53]

1. As we mentioned earlier, we have an explanation of the fact that ergodic systems reach an equilibrium. We know that the microcanonical measure of the set of states that are compatible with equilibrium is large. Therefore, when ergodic systems are concerned, we can be almost certain that the system will spend most of the time in the set of equilibrium states. Furthermore, because the set of nonequilibrium states is small, the fluctuations from equilibrium will be almost certainly rather infrequent on average. This explanation may be used to support the claim that ergodicity implies irreversibility. We expect ergodic systems to evolve irreversibly toward an equilibrium.

---

[52] See *Proc. Nat. Acad. Sci.* (1931).

[53] See, for example, the following statement of Farquhar: "That ergodic theory cannot, however, give a complete solution to the problem of irreversibility is inherent in its nature. It does not seek to define the manner in which a dynamical system approaches thermodynamical equilibrium from a nonequilibrium situation on the removal of constraints, this being the province of master equation and transport theory. It seeks only to establish conditions necessary and sufficient for the system to approach equilibrium or at least to exhibit equilibrium properties. In particular, ergodic theory gives no indication of how quickly systems approach equilibrium, and satisfactory prediction of this is a feature that must be demanded of any successful theory of irreversible behavior. As regards irreversibility, ergodic theory has the characteristics of a mathematical existence theorem. And it possesses, accordingly, both the strength and the weakness of such a theorem. Its strength lies in its generality – it deals with arbitrary macroscopic systems in arbitrary situations; its weakness appears whenever attempts are made to apply it to any specific physical situation" (ibid, p. 6).

2. The first problem with the above-mentioned argument is that ergodicity does not imply anything about the evolution of systems after a finite period of time. More importantly, irreversible behavior may be exhibited by systems that are far from equilibrium. Moreover, there are cases where systems that exhibit irreversible behavior do not evolve toward an equilibrium at all. As we mentioned in the previous chapter, the issue of irreversibility prompted Boltzmann to prove his H-theorem. But, the relations between the H-theorem and the ergodic approach are far from being clear.[54]

3. Another relevant point that was already known in the nineteenth century is that the Stosszahlansatz implies irreversibility. More precisely, it implies an asymmetry between the past and present. It was shown that, if the Stosszahlansatz is true, correlations may disappear either before or after the collisions but not in both cases.

4. There is, however, a more specific issue, one concerning the question of whether Poincaré's recurrence theorem can be reconciled with macroirreversibility, where the ergodic approach yielded a more substantial contribution. It was shown by Mazur and Montroll[55] that the length of the recurrence time is proportional to the number of degrees of freedom and that in the "thermodynamic limit," when the number approaches infinity, genuine irreversibility arises.

Finally, we should say a word or two about the empirical success of the methods of SM. However, we shall address only one aspect of this question. The methods of SM instruct us to treat the evolution of many-particle systems as if it were generated by a stochastic process. Calculations based on this assumption are generally successful. But how can

[54] The H-theorem approach and, more generally, many other results in nonequilibrium SM usually assume some form or another of the assumption of molecular chaos. Using such an assumption, attempts are made to derive kinetic equations without also assuming that the system is approaching an equilibrium. The ergodic approach might provide some justification to this procedure even in this context. See the discussion in Sklar, *Chance and Physics*, ibid, chs. 6 and 7.

[55] Mazur and Montroll, "Poincaré's Cycles, Ergodicity and Irreversibility in Assemblies of Coupled Harmonic Oscillators," *J. Math. Phys.* 1(1) (1960). The article addresses examples where time reversals can be induced by scattering slow neutrons or by magnetic resonance techniques. The result of Mazur and Montroll qualifies as an inductive consistency argument (see Section 2.9). It is only when we assume that we already know that irreversibility exists that we can accept the assumption that we may treat finite systems as if they were actually infinite. In other words, irreversibility is not derived from first principles.

we explain the success given the fact that the evolution is governed by the Hamiltonian and, hence, is completely deterministic?

1. The modern ergodic theory provides us with a plenum of definitions and theorems that characterize stochastic behavior in a way that does not presuppose an underlying indeterminacy. We have already introduced the definition of ergodic systems and mixing. Higher in the hierarchy we find K-systems and Bernoulli systems. Systems of the latter kind have all of the properties of truly stochastic systems, but they can be instantiated by deterministic processes. More precisely, Ornstein and his co-workers established that deterministic systems may be isomorphic to stochastic processes. This last point is quite important. Suppose that we had a proof that established that a given deterministic system was a Bernoulli system. In such a case, no "stochasticity test" would distinguish between this system and, say, a roulette system.[56] In fact, Sinai proved that certain billiard ball systems were isomorphic to Bernoulli systems.

2. In most concrete cases, physicists use the fact that the systems of interest in SM have many degrees of freedom to establish the applicability of Gibbs's rules. The use of such an argument transcends the methods of ergodic theory that aspire to provide us with "purely dynamic" characterizations of stochastic behavior that might apply to any system.

The long list of positive and negative assessments of the ergodic approach that we have compiled in this section should not be construed as a survey. We cannot pretend to have done justice to the deep and complex results that have been obtained in the field. What we are looking for, though, is a way to begin to compose a coherent account. In particular, we want to find out whether the ergodic approach can shed some light on the nature of the probabilities that we encounter in SM. The verdict so far is rather inconclusive. We seem to have encountered several tremendously important ideas and concepts. Nevertheless, some of the most basic questions remain unsolved. Two problems in particular stand out. First, ergodic theory by itself does not solve the problems concerning the interpretation of SM pro-

---

[56] This is a consequence of Ornstein's isomorphism theorem, one of the deepest results in ergodic theory. See Ornstein, *Ergodic Theory, Randomness and Dynamical Systems,* Yale Univ. Press (New Haven, CT, 1974).

babilities. We still do not have a way of presenting the probabilities that we use in this context as physical parameters par excellence. Second, ergodic theoretic considerations cannot help us to rid ourselves of the seemingly arbitrary probabilistic assumptions that are habitually made in discussions on the foundations of statistical mechanics. In particular, it does not allow us to *derive* the assumption of equal a priori probabilities. "What, then, is the status of ergodic theory?" asks Farquhar:

If it be once acknowledged that statistical ideas are inserted into the argument, at some stage does not this reduce the theory to a petito principii? May not the statistical concepts be admitted right at the beginning and, if so, does not the theory become more or less equivalent to the approach by way of representative ensembles? . . . The apologia for ergodic theory runs on the following lines: Although some statistical assumption is required, this assumption is very much more general and consequently very much weaker than the corresponding assumption made in the representative ensemble approach. It is in no way to be compared with the assumption of molecular chaos or the Stosszahlansatz used in early work on the kinetic theory of gases."[57]

There are three problems with this apologia. First, it is not exactly clear to us how Farquhar proposes to measure the strength of the various probabilistic assumptions that one finds in the literature on the foundations of SM. No general method is proposed and defended. Second, even if the statistical assumptions of the ergodic approach are weaker than other assumptions, Farquhar does not tell us how to justify them. In particular, no jusification is given for the choice of sets of measure zero. Third, if probabilities are not measurable physical parameters, akin to force or mass, we need to know what kind of quantities they are. The question of interpretation arises even for relatively "weak" probabilistic concepts. All three of the problems are, surely, philosophical questions of the most annoying kind. Therefore, our insistence that they should be answered is likely to arouse a sense of impatience in the less philosophically inclined reader. Be that as it may, we think that the problems we are preoccupied with are important ones, and in the remainder of this book we shall continue to try to fill the basic gaps in the picture. One strategy that arises in the absence of a physicalist interpretation to SM probabilities is to try to present a variant of the ergodic theory that is consistent with a subjectivist

---

[57] See ibid, pp. 12–13.

outlook. This possibility will be explored presently. The basic probabilistic constructions will be interpreted in terms of degrees of belief. Nevertheless, we shall demonstrate that a considerable amount of ergodic theory may be presented within this framework. The result is a theory where time averages play an important role, but where the calculations of space averages are conceived of as expectation values of various parameters relative to an initial assignment of epistemic probabilities.

2.7. A SUBJECTIVIST VERSION OF THE ERGODIC APPROACH

Subjectivists regard probabilities as propositional attitudes. When Ms. X says that the probability of a heads on the next toss of a coin is p, she expresses, according to the subjectivist, her personal degree of belief. This account stands in clear opposition to the objectivist's construal of probability. According to the objectivist, probability statements express properties of physical systems. For example, the probability of obtaining heads on a toss is, according to the objectivist, a measure of the strength of the disposition of the setup to yield heads. This contrast of opinions is often construed as an argument over the semantics of probability sentences. Many writers believe that the two schools advocate different interpretations of the same concept. This view, however, is mistaken. Discussions on semantic matters have to be conducted within the framework of a fixed scientific theory. The argument between the objectivists and the subjectivists, on the other hand, rages over whether probabilities should be treated as theoretical terms in the first place. The objectivist believes that probabilities have much in common with theoretical terms such as mass or force. The consequential subjectivist, on the other hand, rejects this picture and by so doing undermines the claim that probabilities should be treated as theoretical terms. According to the subjectivist, to say that a state of affairs is objectively probable is an objectionable manner of speech; it involves an improper reification of a propositional attitude.

The subjectivist's critique of objective probability by now should be fairly familiar. The positive vision of the subjectivist, though, is shrouded with much more ambiguity. The ultrasubjectivists insist that one should regard probabilities as extra theoretical concepts that are used to facilitate inferences. Jaynes, for one, attempted to dismantle SM and reformulate it as an inferential scheme rather than as a physical theory.[58] Many subjectivists, though, are much more reluctant to take such a revisionist stand. Many of those who are sympathetic to the subjectivist warning against excessive reification in general, and against the reification of probabilities in particular, will nevertheless hesitate before they approve of the idea of dismantling physical theories that contain probabilities.[59] The ambiguity of the subjectivist position is particularly evident among those who arrived at their subjectivist views as a result of their realization that frequentism cannot furnish a complete account of probabilities. Many writers reject frequentism once they realize that the very definition of frequency depends on the availability of prior probabilistic concepts. This realization often motivates views that aspire to combine subjective probabilities and frequencies. It seems very far afield, though, from a full blown attempt to revise physics.[60]

A particular case in point is the view of those who wish to reformulate the ergodic approach once they realize that the concept of time averages, a concept with close ties to the concept of relative frequencies, is not sufficient as a foundation of probability in the context of SM. We have also seen in this context that additional probabilistic concepts are necessary; we need them to give content to expressions such as "almost everywhere," "almost always," almost certainly," and so on. How should these concepts be interpreted? Some writers who were puzzled by these issues began to consider the possibility of interpreting these additional probabilities as degrees of belief, but many found ultrasubjectivism much too hard to swallow. The questions, then, are

---

[58] See chapter 1.
[59] The reader should note the fact that objectivists also may attempt to dismantle SM and present it as a collection of statistical inference rules. More generally, we should distinguish between (i) the issue of whether SM is a genuine physical theory or merely a collection of inferences, and (ii) the interpretation given to these inferences. This version, though, is problematic for the following reason. Many objectivists believe that the objectivity of statistical inferences stems from the fact that the probabilities reflect the physical properties of the systems involved. The view we alluded to reverses the argument and explains the objectivity of SM in terms of an independently justified objective statistical theory. Because of this problem, we will not pursue this view further.
[60] The same applies to the members of another camp of physicists, namely, those who wish to distinguish between the "objective" probabilities we find in quantum mechanics and those probabilities we find in statistical mechanics. The members of this camp may wish to characterize SM probabilities as epistemic probabilities. Many of them, however, will not be receptive to the idea of dismantling statistical mechanics altogether.

these: Is there a middle ground? Is it possible to introduce epistemic probabilities into physics without accepting Jaynes's extreme position?[61] Even more to the point, one should ask whether it is possible to formulate a coherent subjectivist version of the ergodic approach. This question will occupy the remainder of this chapter.

Let us begin with a few remarks on the subjectivist's attitude toward the very basic concepts of ergodic theory. A clarification of the subjectivist's attitude is absolutely necessary to prepare us for the more advanced discussions that will follow.

1. For a subjectivist, the phase space S of an n-particle system M should be treated as a collection of descriptions of events of the form "M is in a state s." The point s is, in turn, a point in E^{6n}. Taken as a space of events, S is assumed to have a special property:

   P1. *The events of S cannot be further refined.*

   P1 is a consequence of the belief that once we know the exact mechanical state of M, no further relevant and substantially new information on M may be obtained.

2. The σ-algebra B is regarded, as usual, as a descriptive apparatus with certain closure conditions. In our present context, though, B has the following special property:

   P2. *B is descriptively adequate.*

   P2 means that by using predicates from B we can characterize the members of S adequately. More specifically, we require that: (a) For every $s \in S$, there is an $A_s \in B$ such that $s \in A_s$, and if $s' \in A_s$ then s' is a good approximation of s. (b) For every $s,s' \in S$, there are $A_s, A_{s'} \in B$ such that $s \in A_s$, $s' \in A_{s'}$, and $A_s \cap A_{s'} = \emptyset$. (In such a case, of course, $s \notin A_{s'}$ and $s' \notin A_s$.)[62]

   The definition of the notion of approximation requires that we define a distance metric on S. A sufficient condition for a metric to be definable on S is the following: Every collection of open sets whose union covers S has a countable subset that covers S as well.

---

[61] Some readers might have formed the opinion that all we need to do to solve the problems we are discussing is to adopt whatever numbers one finds in physics books as our subjective probabilities. This view is unsatisfactory for the following reason. The problems under consideration are internal to physics and the ones that were discussed, at length, by those statistical physicists who "wrote the numbers." These arguments are not idle exercises; they are important issues that determine the meaning and scope of SM.

[62] These types of requirements are known as separation axioms.

Open sets might be small, but they are always supposed to be appreciable in size.[63]

3. Another special property of S is the following:

   P3. S *can be decomposed into a set of trajectories. Each state lies on exactly one of these trajectories.*

   P3 can be referred to as the "assumption of the deterministic nature of M".[64]

4. Another assumption is that the behavior of M through time is observable. More precisely, we require that

   P4. *The trajectories of* M *can be discretized without a loss of information.*

   The idea here is that a collection of equally spaced points on a trajectory of M, corresponding to a sequence of observations made on M in short and regular intervals, contains essentially all of the information about that trajectory.

5. Finally, S must have the following property:

   P5. *The laws that determine the behavior of* M *remain invariant in time.*

   P5 may be regarded as a strengthening of P1. P1 asserts that when we specify s we have a complete characterization of M. P5 asserts that if we could reproduce s at different times, the resulting evolution whould be identical.

P5 makes it possible to introduce a distinction between closed and open systems. Open systems are systematically influenced by "external" factors. Therefore, the laws that govern their evolution may change when these external disturbances occur. Closed or isolated systems, on the other hand, are not affected by outside forces. Because of that, it is decreed by P5 that when external influences are assumed absent, there are no further degrees of freedom that could produce a time-dependent variation in the development of M.[65]

P1–P5 are special background assumptions on S, which makes it different from more pedestrian spaces of events. Because of these

---

[63] A we remarked earlier, it is often required that S be compact, which means that every open cover has a finite subcover.

[64] In fact, it is also required that each trajectory be parametrized with respect to time.

[65] Notice that without P1–P5 we cannot even formulate the notion of a conserved quantity.

assumptions, S is a "home" of a physical theory rather than a space of events that is constructed in an ad hoc way for specific predictive purposes. One may even go one step further and claim that P1–P5 characterize the canonical form of scientific theories.[66] One should remember, though, that even if there is very little doubt that assumptions P1–P5 are justified in the context of physics, there are plenty of people who argue against making similar assumptions in economics, psychology, sociology, and even biology. It seems that in his early writings de Finetti himself presented his subjectivist outlook as an objection to the positivist tendency to include such assumptions in the social sciences. Indeed, it is not difficult to be sympathetic to the view that one should attempt to avoid restrictive assumptions in the context of, say, economics or psychology whenever possible.[67]

Let us suppose, then, that assumptions P1–P5 are acceptable to the subjectivist. Once they are granted, the subjectivist may consider the possibility of representing the phase spaces of Hamiltonian systems as shift spaces. This fact is important because it offers a clue on how to abstract away some of the physical background of SM and treat it on a par with stochastic processes of other origins. Let us see how this representation is made.

Let $\mathbb{S} = \langle S, \mathrm{B}, m, T \rangle$ be a dynamic system whose origin is a Hamiltonian operating on a phase space. That is, S is a phase space, B is a collection of measurable areas in the phase space, T is an abstract representation of the Hamiltonian, and m is a phase space measure. Let us now define a *new* dynamical system $\mathbb{S}' = \langle S', \mathrm{B}', m', T' \rangle$ in the following way:

1. S' is a set whose elements are the "discretized trajectories" of S.
2. B' is the σ-algebra generated by the cylindrical subsets of S' (relative to a fixed partition of B-measurable sets).
3. $T' : S' \to S'$ is the shift transformation of the elements of S'.
4. m' is the measure generated by assignments of the following kind:
   If $A \in \mathrm{B}$ and $A_{137} \in \mathrm{B}'$ is our proverbial cylindrical set, then

$$m'(A_{137}) = m'(\{\ldots T^{-1}(s), s, T(s), \ldots \mid T^{137}(s) \in A\}) = m(T^{-137}(A)).$$

[66] We are certainly aware of theories that fail to meet some of these requirements. In those cases, though, conceptual problems are usually present.
[67] In these contexts, additional degrees of freedom may be present as the result of volitional factors, for example.

Recall that a shift space measure m is stationary if it is invariant with respect to the shift transformation. A phase measure, on the other hand, is referred to as "incompressible" iff it is invariant with respect to the Hamiltonian. The connection between stationarity and incompressibility is given in the following lemma:

Lemma I. *Let* $\mathbb{S} = \langle S, B, m, T \rangle$ *be a dynamic system whose origin is a Hamiltonian system and* $\mathbb{S}' = \langle S', B', m', T' \rangle$ *a representation of* $\mathbb{S}$ *as a shift space.* m' *is stationary iff* m *is incompressible.*

Proof: *Let* $\mathbb{S} = \langle S, B, m, T \rangle$ *be incompressible. Therefore, for all* $A \in B$,

$$m(A) = m(T(A)).$$

Let

$$A' = \{\{\ldots T^{-1}(s), s, T(s), \ldots\} \mid s \in A\}.$$

We shall prove that $m'(A') = m'(T'(A'))$

$$T'(A') = \{\ldots T'(T'^{-1}(s)), T'(s), T'(T'(s)), \ldots \mid T'(s) \in A\}$$
$$= \{\ldots, s, T(s), T^2(s), \ldots \mid T'(s) \in A\}$$
$$= \{\ldots T^{-1}(s), s, T(s), \ldots \mid s \in T^{-1}(A)\}.$$

Now,

$$m'(T'(A')) = m(T(A))$$

and

$$m'(A') = m(A).$$

But, because m is incompressible, it is also the case that

$$m(A) = m(T(A)).$$

Hence

$$m'(A') = m'(T'(A')),$$

that is, m *is stationary. The other direction of the lemma is immediate as well.*

Because of Lemma I, we may regard the terms stationary and incompressible to be interchangeable. Although the stationarity

101

requirement seems more loaded and objectionable from a subjectivist point of view, Lemma I proves that the subjectivist should be suspicious of the claim that SM measures are necessarily incompressible.

As you may recall, for each $A \in B$ and each trajectory $O_T(s)$ we may define the time average $O_T(s,A)$. Therefore, when the probabilities are defined on the shift space, we may define the notion of the expectation that a randomly chosen trajectory will have particular time averages. This idea will be the basis of an attempt to create a subjectivist version of the ergodic approach to SM. Before we continue, though, let us ascertain that there is nothing objectionable, from a subjectivist's point of view, about the very definition of shift spaces. More specifically, we shall try to find out whether the definition of time averages on the shift space contains presuppositions to which the subjectivist objects. As we shall see, de Finetti made some important remarks about these issues.

De Finetti never objected to the notion of frequencies per se.[68] On the contrary, he recognized that when we are confronted with a well-defined sequence of fixed events, the relative frequencies in this sequence are easily calculable and very useful quantities. So the question of whether the notion of time averages has an acceptable subjectivist meaning depends on whether a subjectivist could regard the sequence $O_T(s) = \{s, T(s) \ldots\}$ for an arbitrary $s \in S$ as a well-defined sequence. There are various reasons why this question is not a trivial one.

1. $O_T(s)$ is an infinite sequence, and there are several passages in de Finetti's writings that suggest that he was sympathetic to finitism and that he regarded relative frequencies as the properties of "arbitrarily long" sequences rather than infinite sequences.[69] This point does not reflect mere pedantry. Recall that the partial sums $\frac{1}{n}\sum_{i=1}^{n}X_A T^i(s)$ of the finite sequences $\{s, T(s), \ldots T^n(s)\}$ may be arbitrarily far from the limiting average $\lim_{n\to\infty}\frac{1}{n}\sum_{i=1}^{n}X_A(T^i(s))$.

[68] For de Finetti on the subject, see his "Foresight, Its Logical Laws Its Subjective Sources," in *Studies in Subjective Probabilities,* ed. Kyburg and Smokler, Krieger (New York, 1980), pp. 73–78. (Original publication 1937.)

[69] Take, for example, the following remark: "Particular attention must be paid to expressing things always in terms of a finite number of instants (which could be taken arbitrarily large) and not in terms of an infinite number." *Theory of Probability*, vol. II, p. 107.

2. Sometimes it seems that de Finetti does allow hypothetically infinite sequences, provided that the hypothetical character of the sequence can be operationally characterized. There are several senses of operational characterization, and all of them have to be satisfied by a truly operationalized definition of the limit of the relative frequencies in infinite sequences:

   a. The elements of the sequence have to be determined experimentally to a reasonable degree of precision.[70]

   b. The frequencies should converge to the limit uniformly.

   c. The law that generates the sequence has to be differentiable.

   These standards of operationability are different but interrelated. None of them is consistently observed in the context of SM.[71]

3. As we remarked, there is additional reason to doubt the usefulness of the time averages. Ergodicity does not give us any indications concerning the *rate* of convergence to the limit.[72]

Those who take these points seriously cannot proceed along the lines of the ergodic approach. Consequently, such subjectivists must accept the ultrasubjectivist position that we discussed at length in the first chapter, namely, they regard SM as an inferential scheme. A second

---

[70] For de Finetti on the issue of precision and the operational meaning he attaches to measurement, see his *Theory of Probability*, v. II, ibid, pp. 289–302. Notice that we refer to the events e that make up the sequence $O_T(s)$ by using the description "e is the result of the nth iteration of the action of T on s." This way of referring to e may be objectionable because, typically, our ability to determine the $T^n(s)$ will diminish exponentially. This means that the approximate determination of the sequence $\{s, T(s), \ldots, T^n(s)\}$ will be impossible, not only for very large n, but also after a relatively short time. So, it might be objectionable to an operationalist such as de Finetti to say that $O_T(s)$ is a well-defined totality even though we cannot determine it experimentally with a reasonable degree of accuracy.

[71] In an early paper called "Differential Laws and the Renunciation of Determinism," de Finetti argued that, to assume that a series is determined by a differential law, the function describing the law must be differentiable. Notice that, in the case of particle systems, the sensitivity to initial conditions that makes it impossible to determine the $T^n(s)$ approximately is to a large extent the result of the nondifferentiable character of the collisions. For a discussion and references, see von Plato, "De Finetti's Earliest Works on the Foundations of Probability," *Erkenntnis* 31 (1989).

[72] See Farquhar's remark: "Thus all the interpretations for which results have been obtained give but formal solutions to the problem of justifying the use of ensembles and of irreversibility, no estimate of relaxation times being obtained" (ibid, p. 29).

camp is comprised of the permissive subjectivists who are able to stomach the notion of time averages and who continue to look for a subjectivist version of the ergodic approach.[73] Throughout the rest of this chapter, we shall follow the thinking of the permissive subjectivist and assume that time averages are *not* antithetical to the spirit of subjectivism. In other words, we shall assume that, for every $s \in S$, $O_T(s)$ is a well-defined totality so that for all $A \in B$, $O_T(s, A)$ is a well-defined quantity that is an "objective" property of $O_T(s)$. That is, we shall assume that $O_T(s, A)$ is as unproblematic as the percentage of black balls in an urn containing a finite population of balls of different colors. If this point is granted, we may continue to describe a subjectivist version of the ergodic approach using as our clue de Finetti's representation theorem.

As we have tried to make clear, it is not usually the case that $O_T(s, A)$ is the same for every $s \in S$. Even where ergodic systems are concerned, there will be some exceptions.[74] The best we can hope for is that $O_T(s, A)$ will be the same for m-almost every $s \in S$. As we have seen, this corresponds to the ergodic case. If, however, the ergodic case applies only under certain conditions, a subjectivist should aspire to provide a *general* account of SM probabilities. Within such an account, the ergodic case should be presented as a special case. (This, in broad strokes, is the methodology that de Finetti employed when he demonstrated that independence is a special case of exchangeability.) As Von Plato pointed out, the ergodic decomposition theorem offers us a clue as to how to provide such an account.[75] In the same article, von Plato also showed how to formulate the general nonergodic case in a subjectivist framework, how to present the ergodic case as a special case, and how to present the analogy with de Finetti's theorem as a mathematical theorem.

---

[73] Strictly speaking, it seems that de Finetti could not have accepted the ergodic approach to SM because of the reasons we have discussed. On the other hand, in his later writing, he showed a remarkable flexibility in accommodating examples of scientific practice within his theoretical framework. In the *Theory of Probability*, there is a section on the Brownian notion, a section on quantum mechanics, and remarks concerning many mathematical results, all of which suggest that he was more interested in putting "warning signs" for conceptual pitfalls than with actually reforming the scientific practice.

[74] Periodic solutions, for example, always exist and they constitute important, but "rare" counterexamples.

[75] See "The Significance of the Ergodic Decomposition Theorem of Stationary Measures for the Interpretation of Probability," ibid.

Theorem III (Krylov and Bogoliubov 1936).[76] *Let* $\langle S, B, T \rangle$ *be an abstract dynamic system. Let* $M_T \langle S, B \rangle$ *be the set of* T-*invariant measures.*

1. $M_T \langle S, B \rangle$ *is a nonempty convex set (that is, if* $m, m' \in M_T \langle S, B \rangle$ *and* a, b *are positive numbers such that* $a + b = 1$, $am + bm' \in M_T \langle S, B \rangle$*).*
2. *The extreme points of* $M_T \langle S, B \rangle$ *are the ergodic measures (that is, if* m *is ergodic and* $m = am' + bm''$, *then either* a *or* b *equals zero.)*

This theorem is known as the *ergodic decomposition theorem*. As a result of this theorem, we can always assume that T-invariant measures exist and every T-invariant measure can be represented as a weighted average of ergodic measures, and uniquely so.[77]

When $\langle S, B, m, T \rangle$ is not ergodic, we can always find nontrivial invariant sets $A \in B$, such that $A = T^{-1}(A)$ and $0 < m(A) < 1$. This means that, although A does not have a full measure, it might nevertheless be the case that if $s \in A$, then for all n, $T^{-n}(s) \in A$. Moreover, one of the consequences of the ergodic theorem is that we can always partition S into invariant sets S_i. When the S_i are maximal [that is, when $S_i \subset A$, $A = T^{-1}(A)$ and $m(A) > m(S_i)$ imply that $m(A) = 1$] we call these invariant sets *ergodic parts*. It is easy to see that a measure that concentrates on an ergodic part is necessarily ergodic.

Let $\{S_i\}$ be a decomposition of S into finite mutually exclusive ergodic parts,[78] and let $C \in B$. Clearly,

$$C = (C \cap S_1) \cup \ldots \cup (C \cap S_n).$$

Let m_i be the ergodic measure that concentrates on S_i. That is, for every $C \in B$, $m_i(C) = m_i(C \cap S_i)$. Let m be an arbitrary T-invariant measure. Because $\{S_i\}$ is a partition, we have

$$m(C) = m(C \cap S_1) + \cdots + m(C \cap S_n).$$

Let

$$d_i = m(S_i).$$

---

[76] See "Les mesures invariantes et transitives dans la mecanique non-linaire," *Mat. Sbornik* (N.S.1) (43) (1936), pp. 707–710.

[77] As von Plato pointed out, the uniqueness is a result of the symplectic structure. See ibid.

[78] When M is bounded and closed, it is compact. Compact systems can always be decomposed into a finite number of ergodic parts.

Observe that $0 \le d_i \le 1$ and that $d_i + \cdots + d_n = 1$.

It is easy to see that

$$m(C) = d_1 m_1(C \cap S_1) + \cdots + d_n m_n(C \cap S_n).$$

Therefore,

$$m(C) = d_1 m_1(C) + \ldots + d_n m_n(C).$$

In other words, if we think of the d_i as "weights," we can get a clear idea of the sense in which m is a weighted average of the m_i. Notice that, because the m_i are ergodic, for all $C \in \mathbf{B}$ and for m_i-almost every $s \in S_i$, we have

$$O_T(s, C \cap S_i) = m_1(C).$$

Hence, we can write:

$$m(C) = d_1 O_T(s, C \cap S_1) + \cdots + d_n O_T(s, C \cap S_n).$$

This expression has an interpretation that is appealing at an intuitive level. Behind it is the idea that when we assign probabilities to the event that M is in a state $s \in C$, we are in fact calculating the subjective expectation of the various time averages.[79] In other words, the various time averages (like the different possible finite frequencies in a sequence of coin tosses) can be seen as objective properties of the various ergodic parts. The d_i are the subjective weights that we attribute to the events $s \in S_i$. This "weighted suspense" is a result of our ignorance.

There is a clear analogy between this interpretation and de Finetti's conception of his representation theorem as a reduction of independence to exchangeability. Ordinarily, we should be expected to hold nonergodic probabilities because of our ignorance with respect to the identity of the "right" ergodic part. Only in extreme cases, when we choose a single i, for which $d_i = 1$ and for all the rest of the i, $d_i = 0$, shall we have an assignment of probabilities that coincides with the relative frequencies. Note that, according to the above-mentioned interpretation, any additional knowledge concerning the preparation of M will usually be considered relevant to the assignment of $m(C)$.

---

[79] In "Insight ...," ibid, p. 74, de Finetti discusses the subjectivist definition of mathematical expectation.

Generally, there is every reason to believe that such knowledge will be relevant to our beliefs concerning the identity of the "right" ergodic part. Therefore, we may be expected to update our degree of belief by using Bayesian conditionalization on the new information. Only in the extreme ergodic cases will additional information become asymptotically irrelevant.[80] This feature of the ergodic subjectivist approach makes it much more palatable for a Bayesian than Jaynes's ultrasubjectivist conception.

Now it is time to introduce a point that the permissive subjectivist has to confront. The ergodic decomposition theorem, the way we presented it, applies only to stationary measures. That is, only when we use a time-invariant probability to assign our probabilities can we represent our probability measure as a subjective expectation of objective frequencies. In other words, it is only when we tacitly assume that our degrees of belief are constrained to be stationary that we may apply the ergodic decomposition theorem. How, then, should we represent nonstationary assignments? Are such assignments incoherent? More generally, what is the status of the stationary assumption? Can we preserve the content of SM if we give up this assumption? The answers to these questions will occupy us for some time.

2.8. STATUS OF THE STATIONARITY ASSUMPTION FROM A SUBJECTIVIST POINT OF VIEW

Let us begin our discussion on the status of the stationarity assumption with a survey of various attempts to prove that the stationary assumption can be derived from first principles.

One argument that is sometimes cited in support of the claim that the stationarity assumption is necessary for the formulation of SM posits that it is empirically incoherent to choose a nonstationary probability in the context of SM.[81] The argument goes as follows. Take a

---

[80] De Finetti writes: "There is in fact a case in which knowledge of the initial state, provided it is sufficiently remote, ceases to have any influence (this is the so-called ergodic case, . . .). This is, however, something that only occurs under specified conditions, and the fact that it often crops up is merely an indication that these conditions are satisfied in many of the problems one considers, rather than because of some principle which permits us to use the idea indiscriminately." See *Theory of Probability*, vol. II, pp. 149–150.

[81] This seems to be the sentiment expressed by von Plato when he writes: "De Finetti's claim about the unnecessity of objective probabilities, would concern the probabilities of statistically stable phenomena in the strong sense of unique limiting

closed mechanical system and prepare it in a certain way. Next, repeat the preparation at a later time. If you succeed in making the preparation identical each time, the system will have identical properties in both cases. If, in both cases, the system is prepared so that its parameters lie within the same range of values, the probability that it will have certain properties should be identical; that is, the probabilities must be invariant with respect to the time of preparation. In that respect, the time-invariance requirement is a generalization of the assumption of mechanical determinism. The exact specification of the physical parameters determines the subsequent *states* uniquely. An interval specification determines the *probabilities* of the subsequent states uniquely. If the preparation is identical, the argument goes, and there are no "hidden variables," the probabilities must be the same too.

A very stubborn subjectivist might try to avoid this conclusion by maintaining that, even if it looks as if his probabilities were chosen in accordance with the constraints placed by physics, this appearance is merely an illusion based on coincidence. More generally, it is certainly possible to feel that the above argument simply begs the question: Why should we believe that the interval specification is sufficient for determining the probabilities uniquely, independent of the time of preparation? Is there an independent reason for this belief that is more than a reiteration of a prior objectivist conviction?

The second argument that attempts to establish the mandatory nature of the stationarity constraint is rather straightforward. We have seen that the action of a Hamiltonian on E^{6n} preserves the Lebesgue measure on E^{6n}. This is precisely what Liouville's theorem establishes. This theorem, then, certainly sounds like a good reason to believe that, at least where Hamiltonian systems are concerned, stationarity is simply imposed on us![82] There are, however, two reasons why Liouville's theorem does not remove the issue completely. First, Liouville's theorem applies only to conservative systems. However, because most of the systems that one encounters in practice are not completely

---

behavior. But as has been shown above, there exists a physical criterion for this within the theory of dynamical systems" (ibid, p. 429). According to the same logic, the physical criterion for stationarity is whether the physical makeup of the system stays constant over time.

[82] For an article that uses this argument implicitly, see Malament and Zabell, "Why Do the Ergodic Averages Work?", *Phil. Sci.* (1983). On p. 345, the writers assume the invariance with respect to diffeomorphisms, which is the result of the Liouville theorem.

closed, one hopes that it would be possible to find other justifications for the stationarity constraint. More important, though, from a subjectivist's point of view is the fact that the Lebesgue measure does not have a privileged status.[83] It is merely one measure among many. Even the requirement that we choose a measure absolutely continuous with respect to the Lebesgue measure is not acceptable from a subjectivist's point of view.[84] In the eye of the subjectivist, then, the second argument is reduced to an issue to which we shall return again and again, namely, Is there a physical reason that should motivate us to choose only measures that are absolutely continuous with respect to the volume measure?

The third argument we will consider is an attempt to give stationarity a subjectivist interpretation.[85] The motivating idea is simple, but attractive: What if we interpret stationarity as a property of agents and not of roulette tables or mechanical systems? What if we define a stationary or tenacious agent as an agent whose probabilistic opinions do not change over time? Perhaps a subjectivist statistical mechanics can be developed along these lines. The objections to this view, though, are formidable. We have already pointed out that the concept of stationarity was originally developed as an abstract representation of the notion of an incompressible fluid – a particle system whose volume does not change over time. This, of course, does not mean that it is the only way to motivate the concept of stationarity. In particular, the choice of a stationary measure can be regarded as the product of a tenacity constraint, an attitude that characterizes the ratiocination of agents who will not change their degree of belief unless they learn a new piece of relevant information. However, the merits of such an interpretation are questionable. Tenacity requirements are idiosyncratic at best, from a subjectivist's point of view, because subjectivists regard only coherence requirements as being universally binding. Besides, there is much to be said against tenacity from a subjectivist's point of view. The tenacity requirement mandates that, in the case of a conflict between our future beliefs and our present ones, we should

---

[83] If we could justify the choice of the Lebesgue measure as our probability measure, stationarity would follow from Liouville's theorem.

[84] If the requirement of absolute continuity were justifiable, stationarity would follow, again, by Liouville's theorem.

[85] See Skyrms, *Pragmatics and Empiricism*, Yale Univ. Press (New Haven, CT, 1980). The third argument is an application of Skyrms's ideas on SM. Skyrms, however, did not suggest this application; in fact, he does not believe that it applies to the case of SM (private communication).

always correct the former rather than the latter.[86] Therefore, we cannot simply view it as a matter of course that physicists will always respect the tenacity requirement. But, the real problem with the proposal is that tenacity and stationarity reflect different attitudes, even in the context of SM. This difference becomes clear with the following example. Knowing that a target may expand or shrink may be reason to believe that the probability of a dart-throwing device hitting the bull's eye is not stationary. Such knowledge, however, is not reason to give up tenacity. Only the knowledge that shrinking or expansion actually occurred or will occur should induce the tenacious agent to rethink his degrees of belief. The converse is also true. An agent may decide to stop being tenacious for reasons that have nothing to do with his beliefs concerning the constancy of the dispositions of the system. Tenacity, after all, is not a coherence requirement. Therefore, while tenacity is undoubtedly a "natural" constraint on the choice of degrees of belief, it cannot replace or express the insight behind stationarity. A belief in stationarity is motivated by the idea that when we encounter a system whose dispositions do not change in time, we should choose probabilities that also do not change in time. We accept the stationarity constraint only when we believe that the systems with which we are dealing will consistently respond in a like manner to the conditions to which they are subjected. This belief, however, when interpreted probabilistically, is regarded by the subjectivist as an opinion rather than a universal constraint.[87]

---

[86] We may discover that at time t we held $p(A) = p$, at $t' > t$ we held $p(B) = p'$ (p does not equal p'), and at $t'' > t'$ we found out that $A = B$. In such a case we are committed at t'' by the coherence requirement to change either $p(A)$ or $p(B)$; we are not necessarily committed to change $p(B)$. It seems that a tenacious agent would say that because, unbeknownst to him, the formation of $p(B)$ at t' has already violated the tenacity requirement, then, he *has* to change $p(B)$. The principle behind this reasoning is not particularly compelling, and we do not think that it is a consequence of the general spirit of subjectivism. For a discussion of some related issues, see Levi, *The Enterprise of Knowledge*, ibid, section 4.3.

[87] The choice of stationary probabilities usually reflects a belief that the system has fixed dispositional properties. Now systems with fixed dispositions are common, but we are acquainted with many examples of systems whose dispositions are not fixed; the shrinking target is one such system. Therefore, it seems reasonable to expect that the difference between these two types of systems will be reflected in the choice of probabilities. This difference, however, should not be confused with the difference between tenacity and nontenacity that characterizes the attitude of agents to their own beliefs. In other words, it should not be confused with the distinction between agents who choose to accept a principle of epistemic inertia and those who do not. Only when the very concept of a disposition is rejected is it con-

110

The fourth argument is, again, an attempt to give stationarity a subjectivist interpretation. The argument is more complex, but it fails nevertheless. According to a theorem that has been proven by Ryll–Nardzewski [88] and that has received considerable attention, the requirement that a probability measure be exchangeable is essentially equivalent to a stationary requirement. Why, then, can we not say that in the context of statistical mechanics we tend to choose exchangeable probabilities; that is, we tend to disregard the order of the events we observe and instead concentrate on the relative frequencies?

In a nutshell, the answer is as follows: Although exchangeability is a property of the probabilities that we use in many contexts, there are some important contexts in which stationarity arises and exchangeability is not acceptable. Statistical mechanics is one of these contexts. As we have seen, one of the fundamental assumptions of statistical mechanics states that the phase space of mechanical systems is divisible into an exclusive and exhaustive set of trajectories – those trajectories that it is possible for the mechanical system to follow. This assumption has an objectivist flavor, but without it we cannot adequately formulate SM. Once this assumption is granted, however, exchangeability is ruled out.

Let $O_T(s)$ be a possible trajectory for M. Let $w_0, w_1, w_2, w_3, \ldots$ be points on $O_T(s)$ that represent the states that M occupies in consecutive points of time. Let A_1, A_2 be a pair of exclusive and exhaustive properties of a mechanical system. Suppose that we have observed that:

$$w_0 \in A_2, \quad w_1 \in A_2, \quad w_2 \in A_1, \quad w_3 \in A_1.$$

If we regarded mechanical events as exchangeable, then, given the fact that the above sequence occurred, we would be willing to draw the same predictions as those we would have drawn if we had observed, for example, that $w_0 \in A_1, w_1 \in A_2, w_2 \in A_1, w_3 \in A_2$. Such a view, however, does not cohere very well with the basic assumptions that we

---

ceivable to attempt to construe stationarity using tenacity. However, we believe that, whether or not in SM we have to require stationarity, an issue that we shall return to shortly, there are many systems whose dispositions are not stationary, and the differences between the two types of systems cannot reflect only the different attitudes of their observers.

[88] See Ryll–Nardzewski, "On Stationary Sequences of Random Variables and De Finetti's Equivalence," *Colloq. Math.* 4 (1957), pp. 149–156.

have already mentioned. To decide whether, say, $w_4 \in A_1$ or not, we need to inquire whether the trajectory $w_0, w_1, w_2, w_3, w_4, \ldots$ will pass through A_1 or not. There is no reason to expect that the answer will be the same for two exchangeable trajectories.[89] So, in the context of SM, even subjectivists are not expected to choose exchangeable probabilities. As a result, the Ryll–Nardzewski theorem does not supply us with a subjectivist account of stationarity.

This last point shows why the physical background of SM necessitates a genuinely different type of subjectivism. In many contexts, subjectivists use exchangeability as a constraint that partially determines the assignment of probabilities to the possible sequences of events. In the context of SM, subjectivists cannot rely on this method.

There may be other strategies that subjectivists may employ to reinterpret stationarity, but the ones that we have mentioned are the only ones that appear in the literature. In the absence of new ideas, we are left with the more difficult task of exploring whether it is possible to do away with stationarity altogether. To answer this question we will inquire whether (or under what conditions), given a subjective probability measure m, it is possible to find another probability measure m' such that m' is stationary and m is equivalent to m'.[90] (It is important to note that the notion of equivalence does not imply a total coincidence of the betting behavior. The equivalence relation implies only that, where limit operations are concerned, different equivalent measures will yield the same results. In the context of ergodic theory, though, where infinite time averages are the most important concepts, the notion of stochastic equivalence is the most natural notion of equivalence.)

Definition. *Let* $\langle S, B, T \rangle$ *be a dynamic system and* m $\in$ M$\langle S, B \rangle$ (m *is not necessarily T-invariant*). m *will be called a nondissipative measure*[91] *if, for all* A $\in$ B, *such that* m(A) > 0,

---

[89] From a dynamic point of view, there is no particular relation that holds between exchangeable trajectories.

[90] For an article on the subject, see Hurewicz, "Ergodic Theorem Without Invariant Measures," *Ann. Math.* 45(1) (1944), which, to the best of our knowledge, is the first article on the subject. See also Oxtoby, "On the Ergodic Theorem of Hurewicz," *Ann. Math.* 49 (1948); Dowker, "Invariant Measures and the Ergodic Theorem," *Duke Math. J.* (1949); and Dowker, "Finite and Sigma Finite Measures," *Ann. Math.* 54(3) (1951).

[91] For a different but related definition of nondissipativeness, see Hurewicz, ibid.

$$\lim\inf_{n\to\infty} m(T^n(A)) > 0.$$

Theorem IV (Dowker 1954). *Let* $\langle S,B,T \rangle$ *be a dynamic system. Let* m $\in$ $M\langle S,B \rangle$ *(m is not necessarily T-invariant). A necessary and sufficient condition that there exists a finite T-invariant measure* m', *such that* m' *is equivalent to* m, *is that* m *is a nondissipative measure.*

The idea behind the proof is that if m is nondissipative, the quantities

$$m_n(A) = \frac{1}{n}\sum_{i=1} m(T^i(A))$$

converge in the generalized Banach–Mazur sense.[92] Let $m^* = \text{Lim}\, m_n$, where Lim is the Banach–Mazur limit. First, it can be proven that m^* is a measure. The fact that m^* is equivalent to m is an easy consequence of the definitions: If m is nondissipative, $m_n \sim m$ for all n; from the properties of the Banach–Mazur limit, it follows that Lim $m_n \sim m$ as well. Next, it can be shown that m^* is T-invariant. To see why, observe that

$$m^*(T(A)) = \text{Lim}\, m_{n+1}(A) = \text{Lim}\, m_n(A) = m^*(A).$$

The theorem proves that the crucial issue for the subjectivists is not stationarity but nondissipativeness. Strictly speaking, a subjectivist can claim that the theorem shows how to represent stationary measures as special averages of the measures m_n.[93] However, this does not solve the problem of the subjectivist.

---

[92] For the definition and properties of the Banach–Mazur integral, see the appendix by Banach to Saks, *Theory of the Integral*, and his "Theorie des Operatione Lineaires," *Monografji Math.* 1 (1932). The Banach–Mazur limit is not always uniquely defined for a sequence $s1, s2, \ldots$, but it is always possible to choose $\lim \inf\{s_n\} \le \text{Lim}\{s_n\} \le \lim\sup\{s_n\}$. When the ordinary limit $\lim\{s_n\}$ exists, $\text{Lim}\{s_n\} = \lim\{s_n\}$.

[93] As for the relations between time averages and relative frequencies in the nonstationary but nondissipative case, the following considerations are quite illuminating. In the stationary ergodic case, we saw that for all $A, A' \in$ B,

$$\lim_{n\to\infty} \frac{1}{n}\sum_{i=1}^{n} m(T^i(A) \cap A') = m(A)m(A').$$

Therefore,

113

The notion of dissipativeness has an important physical background that makes it even harder to claim that we can choose our probabilistic assignments freely in the context of SM. Dissipative systems lose energy to their surroundings, and hence we cannot consider them as

$$m(A') = \frac{1}{m(A)} \left(\lim_{n \to \infty} \frac{1}{n} \sum_{i=1}^{n} m(T^i(A) \cap A') \right).$$

Because $m(A)$ is a constant, we can write:

$$m(A') = \lim_{n \to \infty} \frac{1}{n} \sum_{i=1}^{n} \frac{1}{m(A)} (m(T^i(A) \cap A')).$$

And because in the stationary case for all i, $m(A) = m(T^i(A))$, it is also the case that:

$$m(A') = \lim_{n \to \infty} \frac{1}{n} \sum_{i=1}^{n} \frac{1}{m(T^1(A))} (m(T^i(A) \cap A')).$$

As we have seen, this equation may be given a rather suggestive interpretation. If we want to measure A', we may do so by choosing an arbitrary "test set" A and calculating the average of the overlaps $m(T^i(A) \cap A')$ weighted by $1/m(T^i(A))$. As you may recall, when m is ergodic, this way of measuring $m(A')$ is independent of the choice of A. In fact, as long as T is ergodic, it is independent of the choice of T as well. Observe that:

$$1/m(T^i(A)) (m(T^i(A) \ll A')) = m(A'|T^i(A)),$$

where $m(A'/T^i(A))$ is the measure of A' conditional on $T^i(A)$. So this way of measuring $m(A')$ can be seen as the averaging of the proportion of the A' in n "random" samples. Note that the same idea may apply even when m is not stationary, provided that it is a nondissipative measure. The iterates $T^i(A)$ of the "test set" A may "expand" or "shrink." But because we measure $1/m(T^i(A))(m(T^i(A) \cap A'))$, the division by $1/m(T^i(A))$ compensates for the changes of size when the overlaps $m(T^i(A) \cap A')$ are measured. Observe that when we choose a test set A with $m(A) > 0$, if m is nondissipative, the expression $1/m(T^i(A)) (m(T^i(A) \cap A')$ will almost always be well defined because $m(T^i(A)) = 0$ only for a set of indexes i with density zero. This construction, then, proves that, even in the nonstationary but nondissipative case, there is a conceptual relation between ergodicity and time averages. This connection yields a more general relation between ergodicity and relative frequencies as well.

Theorem V (Hurewicz 1943). *Let $\langle S, B, T, m \rangle$ be a nondissipative ergodic system. For m-almost every $s \in S$, and every $A \in B$,*

$$m(A) = O_{T,f}(s, A) = \lim_{n \to \infty} \frac{1}{n} \sum_{i=1}^{n} f(s) X_A (T^i(s)),$$

where f is a square-integrable m-measurable "generalized average" function that is independent of the choice of s and A.

114

closed systems. However, the formulation of a nonstationary ergodic theory does lead to an important generalization of the concept of relative frequencies that may be incorporated within the subjectivist framework.

2.9. SM AS AN INDUCTIVE CONSISTENCY ARGUMENT

In the previous section we criticized various attempts to construe stationarity as a constraint that is imposed on our degrees of belief in the context of SM. The grounds for accepting such a constraint, we argued, are questionable from a subjectivist's point of view. The objections do not only apply to an explicit acceptance of the stationarity assumption. After all, one may attempt to derive stationarity from other constraints. But the result is that objectivist assumptions are invariably smuggled in. For example, one may attempt to derive stationarity from the principle that when the dispositions of a system remain fixed our degrees of belief should also remain the same. This principle, however, is acceptable only to those who believe that the probabilities themselves reflect the objective dispositions of systems. Therefore, this belief is one of the central theses that are attacked by subjectivists. Subjectivists believe that we may change our degrees of belief in the face of a new piece of information, even one that has no direct relation to the system at hand.

Behind the specific points of criticism, though, there is a more profound problem. What the permissive subjectivist has failed to do, at least so far, is to provide a new outlook on the nature of probabilities. The objectivist regards SM probabilities as theoretical entities that are internal to physics. Ultrasubjectivists place probabilities outside physics and regard them as tools of inference. What then is the vision of the permissive subjectivist? Is there a third position that is genuinely different from the two already discussed? Can the permissive subjectivist avoid the danger of sliding into either extreme?

Our belief is that there is a third position, one that was more

---

Notice that, as long as the process of the generalized averaging is fixed, the generalized average $O_{T,f}(s,A)$ can be seen as an "objective" property of $O_T(s)$ that is as unproblematic as $O_T(s,A)$ from a subjectivist point of view. In that sense, the generalized averages do not introduce any additional problems to the subjectivist formulation of the ergodic approach.

115

common among physicists at the end of the nineteenth century than it is today. Nevertheless, it is a consistent position that deserves to be articulated in some length. According to this position, probabilities may express relations *between* physical theories. Confronted with theories A and B, we may ask ourselves how credible is theory A relative to theory B (that is, were theory B proven to be true)? More specifically, we may look for a demonstration that the acceptance of theory B will not make theory A vastly improbable. We shall call such a demonstration a *proof of the inductive consistency* of A and B. In the context of our discussion we may attempt to show that probabilities are introduced into physics to prove the inductive consistency of mechanics and thermodynamics. We believe that this picture could provide a plausible background for the permissive subjectivist's construal of SM. We also believe that the inductive consistency approach to SM is a genuinely intermediate position. It shares with the ultrasubjectivist view the rejection of the idea that probabilities are to be treated as theoretical entities. Nevertheless, it accords to probabilistic reasoning a function that transcends its inferential role. This position will be explored in this section.

The history of the third position to SM began more or less at the end of the last century. Bernard Bunhes, the director of the Puy-de-Dome Observatory, for example, must have had such a view in mind when he wrote that "the most important question, perhaps, of contemporary scientific philosophy is that of the *compatibility or the incompatibility of thermodynamics and mechanism*"[94] (italics added). The specific problem that Brunhes had in mind was that the irreversibility of the second law of thermodynamics seemed to contradict the reversibility of Newtonian physics. Consequently, the task of SM, according to Brunhes, was to establish the *consistency* of these two physical theories. Brunhes's formulation of the issue is to be distinguished from the more extreme views of his contemporaries, such as Boltzmann, on the one hand, who believed that SM, augmented by the assumption of molecular chaos, could replace thermodynamics altogether. Within such a framework, Boltzmann proved the *H*-theorem, which provided an explanation, in terms of molecular motion, of the emergence of irreversibility. The other extreme view was held by Planck, his student Zermelo, and a host of others. The members of this

---

[94] See Brush, *Statistical Physics*, p. 96.

camp maintained that the issue of irreversibility proves that physicists ought to give up the Newtonian picture.[95]

The existence of these three camps was, of course, completely natural. If we encounter two theories, A and B, both of which seem to be true but that seem to be nevertheless inconsistent with one another, we have three different courses of action open for us. The first and the second both amount to giving up either of these two theories. These courses of action always involve a process of salvaging what can be, and should be, salvaged from the discarded theories. The third course of action is to devise an argument proving that, contrary to appearances, A and B can both be consistently held. This course of action often involves a discussion on a correspondence principle that is to guide us when we move from "the domain of A" to that of B. The idea behind such principles is to minimize the dangers created by the potential inconsistencies between the theories. There are quite a few examples of the use of these strategies in physics.[96] Nevertheless, there is almost no discussion on these cases in the literature on the philosophy of science. Therefore, it may be beneficial to begin with a few general remarks.

A proof of the consistency of theories A and B may take one of three forms: deductive, inductive, or informal. A deductive consistency proof attempts to establish that the inconsistency between A and B is only apparent. Two consistent theories that are deductively consistent with each other cannot have contradictory consequences. An inductive consistency proof of A and B, on the other hand, is possible even if A and B are, strictly speaking, inconsistent. Take, for example, a case where A implies that "all P's are Q's" and B implies that "some

---

[95] One should note, though, that according to Planck irreversibility was to be explained in terms of the theory of radiation. More generally, Planck's antagonism toward the mechanical conception of the world was the result of his view that, ultimately, all physical laws are to be formulated in terms of the relations between attractive and repulsive forces. This view, which was known at the time as energeticism, was not shared by all of the physicists who were antagonistic to the mechanical world view. As we mentioned earlier, Mach, for example, objected to Boltzmann's physics on the grounds that it violated the principle of the economy of thought.

[96] Quantum theory offers many such examples. In particular, we believe that Bohr's notion of complementarity may be analyzed as an informal consistency proof of classical and quantum physics. Daneri, Prosperi, and Loringer tried to provide a more rigorous inductive consistency proof [see *Nuc. Phys.* 33 (1962), pp. 297–319.] Also see my "Notes on the Measurement Problem" (forthcoming.)

117

P's are not Q's." A and B are, of course, inconsistent. Nevertheless, they may be inductively consistent if the probability of obtaining a P that is not a Q is very small. More generally, an inductive consistency proof is an argument that A does not deem B improbable. Therefore, inductive consistency proofs usually involve statistical arguments. The third type of consistency proof is much more difficult to define rigorously. Informal consistency between theories means that one may avoid the dangers of the inconsistencies by following a set of informal rules that tell us how to proceed in every case. (These rules may or may not be accompanied by further explanations.) This is, for example, how the inconsistencies between the relativistic theory of gravity and quantum field theory are often treated. Because the notion of informal consistency will not play any role in our subsequent discussion, we shall not attempt to explain it any further; instead, we shall make do with making just one observation. Possessing an algorithm that provides us, in every eventuality, with a unique and unambiguous recommendation for a course of action may solve most of the *practical* problems created by the inconsistency between A and B. Hence we may regard this case as a very weak case of consistency between theories.

In the case of thermodynamics and mechanics, it is not possible to find a proof of the deductive consistency of the theories. For example, we know that periodic motions will contradict the second law of thermodynamics. Nevertheless, we may attempt to prove that such motions are extremely improbable. Such a proof establishing the inductive consistency of the two theories will demonstrate that we need not give up either thermodynamics or Newtonian mechanics. Note that a general proof of the inductive consistency of thermodynamics and mechanics is an extremely difficult matter. We know, for example, that at low temperatures liquids solidify, certain metals become magnetic, and certain crystals are formed. However, it is quite difficult to prove that these processes, which give rise to very fine structures, might be described as the outcome of the motion of an incredibly large number of irregularly moving particles. But it is a much harder task to argue that it is "overwhelmingly likely" that trillions and trillions of particles will invariably settle in those fine formations that yield magnets and crystals. What we need is a proof that those structures whose "spontaneous creation" appears miraculous are created spontaneously all of the time. Only such a proof will establish that thermodynamics is not deemed vastly improbable by the basic laws of mechanics. It was not until 1925 that

118

Ising[97] made the first step toward a solution of the problem. His ideas were developed by Onsager and others to a fairly satisfactory theory of the emergence of new and qualitatively different macroscopic properties. We shall return to the discussion of these ideas later in this chapter.

Another, much easier aspect of the proof of the consistency of mechanics and thermodynamics concerns the behavior of enclosed gas systems. In fact, we have already given an implicit inductive consistency proof that is appropriate for the case of an n-particle "ideal gas" system. The proof can be obtained from considerations introduced by Birkhoff and Khinchin. The latter showed that many-particle systems are ergodic, and the former proved that ergodic systems do not have nontrivial invariant parts. This last fact implies that the set of periodic motions, for example, a set that is clearly invariant, has to be a negligible set. These two points taken together constitute an inductive consistency proof. As we have seen before, the proof can be generalized considerably.

Note that when our aim is to provide an inductive consistency proof we do not have to justify our choice of probabilistic assignments. Only when we try to *derive* the approach to equilibrium do we have to avoid unsubstantiated assumptions. In our present frame of mind, we may assume that we already know that thermodynamics is a true theory. All we need to establish is that the laws of mechanics do not deem the second law of thermodynamics highly improbable. To that end, it is sufficient to argue that the conclusion follows relative to *some* probabilistic assignments on the phase space.[98] In particular, we may restrict our attention to the set of stationary measures $M_T \langle S, B \rangle$.[99] This conception of SM makes its aims relatively easy to achieve. Nevertheless, one must not forget that inductive consistency proofs do not yield an independent support to either theory. The proof we outlined, for

[97] For further reading on the subject, see Brush, *Statistical Physics*, pp. 239–250.

[98] Note that we assume, as before, that A and B do not conflict with any other aspect of our knowledge. Hence we take it that the burden of truth is not on proving that the theories are correct, but on those who claim that either theory should be discarded. It is because of this shift that we may choose any consistent probabilistic assignment and construct an inductive consistency proof relative to it. (Of course, a choice of a natural assignment of probabilities will make the argument more convincing.)

[99] Of course, if we find an inductive consistency proof relative to a wider set of measures, the resulting proof will be more significant from a foundational point of view.

example, merely indicates that the fact that we expect an equilibrium is consistent with the laws of physics. There is no attempt to reduce the laws of thermodynamics to mechanical laws or even to replace them with an inferential scheme. We should regard such proofs as analytic devices designed to convince ourselves of the inductive consistency of two sets of beliefs whose truths were never seriously doubted. In other words, the proof is concerned with the structure of our beliefs rather than with features of the objective world.[100] Indeed, most of the ergodicity proofs that we possess are so abstract and sketchy that it is not too difficult to see why one may insist on taking them with a grain of salt.[101]

---

[100] Another formulation of the aims and methods of the inductive consistency proof is the following. Let A and B be two contradictory theories. Let $M_{IC}\langle A,B \rangle$ be the set of measures relative to which an inductive consistency proof of A and B may be given. Assume that $M_{IC}\langle A,B \rangle$ is not empty. The question, then, is whether there are background assumptions that prohibit the choice of all of the members of $M_{IC}\langle A,B \rangle$ as our probabilistic assignments. If there are such background assumptions, A and B are not inductively consistent. If at least some members of $M_{IC}\langle A,B \rangle$ cohere with the physical background, we may claim that A and B may be regarded as inductively consistent. To illustrate a case where none of the members of $M_{IC}\langle A,B \rangle$ cohere with the background assumptions, take the following example. Suppose that we had a mistaken belief that even dissipative systems always reach an equilibrium. In such a case, we would try to find a probabilistic assumption on the phase space of the dissipative system relative to which an equilibrium will be reached invariably. Now we know that, after the process of dissipation takes its course, the system will reach an area of the phase space that is a proper subset of the original space; call this area E. It is easy to see that E is an invariant subset. So, clearly, if the motion is stationary, it cannot be ergodic. In many cases, E is a hyperspace of lower dimensionality. In such cases, the motion is either nonincompressible or nonstationary. (Indeed it is not! That is to be expected when dissipative systems are concerned!) So if all of the members of $M_{IC}\langle A,B \rangle$ relative to which an inductive consistency proof can be given are stationary (or at least have some weaker stationarity properties), none of them will be permitted to be chosen because they all conflict with the physical background. This example, then, establishes the well-known fact that the theory of dissipative systems is not inductively consistent with the laws that govern the motion of conservative systems.

[101] In quantum SM, the assumptions needed for establishing an ergodicity proof are even more objectionable than in the classical case. First, the assumption of the nondegeneracy of the energy eigenvalues is not a realistic assumption (see Jancel, ibid, p. 65). In addition, we need to reconcile the fact that in the microlevel we expect to encounter noncommuting operators while in the macrolevel such operators are completely absent [von Neumann emphasized this point in Z. *Physik* 57, 30 (1929)]. The solution to this problem requires, again, that we refer to the facts of quantum thermodynamics as grounds for accepting the necessary assumptions. This prevents the "quantum ergodic theorem" from being a proof from first principles, and consequently it does not yield a reduction of thermodynamics or, more generally, any independent support to the thermodynamic laws.

Let us end the discussion on inductive consistency proofs with a further remark. In many books on SM it is stated that ergodic systems have no nontrivial constants of the motion. This means that, even if we know some nontrivial information about the way the system was prepared, we may discard such information as being asymptotically irrelevant. This feature of the characterization of ergodicity is rather problematic in inferential contexts. In fact, it was precisely because of these problems that we believe that the ultrasubjectivist idea of regarding SM as an inferential scheme is problematic. If, on the other hand, probabilities are introduced into SM solely for the purposes of proving the inductive consistency of thermodynamics and mechanics, we need not worry about this problem. Even if we think about probabilities as degrees of belief, we do not use them to infer anything; we use them to prove to ourselves that our scientific theories "hang together." In such a context, the nonexistence of nontrivial invariants of the motion has a new and natural interpretation. It signifies the fact that we may choose any coherent probabilistic assignment relative to which an inductive consistency proof can be constructed. When ergodic systems are concerned, no "hidden constraints" are lurking in the background.

2.10. THE ISING–LENZ THEORY AS AN EXAMPLE OF AN INDUCTIVE CONSISTENCY ARGUMENT

Some readers might have formed the opinion that theories that are based on inductive consistency arguments are barely acceptable, second class citizens that one should be ashamed of or ersatz theories that should be replaced whenever possible. In this section we would like to correct this picture. We shall introduce the Ising–Lenz model of phase transitions, which is a very powerful and indispensable theory that is based on an inductive consistency argument. As we shall see, the theory depends on a premise that is dissonant with the underlying mechanical framework. Hence the theory cannot be construed as a reduction of a thermodynamic law to more basic mechanical principles. However, the Ising–Lenz theory is far more informative than any other competing phenomenological theory. This important point demonstrates that theories that are based on inductive consistency arguments might nevertheless be extremely valuable. They might give rise to explanations and predictions that cannot be obtained in any other way. More generally, the fact that a theory depends on an inductive consistency argumency does not mean that it cannot be important and useful; it often happens that once such theories are accepted they generate new and powerful results that are taken with complete seriousness.

Let us begin with the presentation of the Ising–Lenz theory of phase transitions.[102] The presentation will be completely informal and rather sketchy at times. However, we feel that there is a general need to present the theory of phase transitions to the general philosophical community, because an understanding of this theory is vital to those who wish to discuss the notion of the emergence of properties, the notion of supervenience, and the issue of reduction of scientific theories. The reader should keep in mind that in the entire Hall of Science one cannot find a more ingenious and thorough attempt to discuss these issues in a rigorous manner. Our first step is the presentation of the basic mathematical framework. We shall begin with a more general and abstract discussion that will be followed by a concrete example, namely, the theory of ferromagnets.[103]

The basic mathematical object with which we shall be preoccupied is a regular two-dimensional square *lattice*.[104] The intersections of the lattice lines are called *sites*. For example, when we model ferromagnetic phenomena, the sites of the lattice represent point particles with a magnetic moment called *spin*. (It is usually assumed that the set of sites is denumerable; hence we may designate them as $\{\sigma_i\}_{i \in N}$. Each site can be in various *positions*. The spin, for example, can have two values: It is either "up" or "down." We represent the positions by assigning to each site σ_i the values "1" or "−1," respectively.

When all of the sites are assigned positions, we obtain a well-defined lattice *state*. For example, when all of the point particles are assigned with spins having a definite direction, the system is in a well-defined magnetic state. It is often assumed that the lattice is homo-

---

[102] For a more detailed account of the theory of phase transitions, see, for example, Robert B. Griffiths, "Rigorous Results and Theorems" in *Phase Transitions and Critical Phenomena*, vol. 1, eds. Domb and Green, Academic Press (London, 1972), pp. 8–109. For an introduction, see David Chandler, *Introduction to Modern Statistical Mechanics*, Oxford Univ. Press (New York, 1987), pp. 119–159.

[103] We shall also mention the application of Ising models to the liquid–solid transitions.

[104] Regularity means that the lattice is composed of a system of squares. Most of the proofs and arguments that we shall cite depend on the assumption that we are dealing with regular two-dimensional lattices. In some cases, though, weaker assumptions can be used instead.

geneous, which means that all of the sites are "qualitatively" the same.

Next we have to discuss the law that governs the evolution of the system. We will do so by specifying the Hamiltonian of the system, which, in our case, will assume a particularly simple form:

$$\mathbf{H} = -J\sum \sigma_i \sigma_j + H\sum \sigma_i$$

where J and H are constants and the summation $\sum \sigma_i\sigma_j$ is taken over nearest neighbors only. (The assumption is that we need to take into account the potential between nearest neighbors only. Each pair is counted only once.) When J is positive, like spins attract each other.[105] This is the ferromagnetic case. When J is negative, we obtain the anti-ferromagnetic case where opposite spins attract each other. The constant H represents the effect of an external magnetic field. (In much of the discussion that follows it will be assumed that $H = 0$). The reader should note that the idea of restricting our attention only to nearest-neighbor interactions requires a strong assumption that is not made as a matter of course. In general, we assume that the influence of one site on another wanes in proportion to their distance. In our example, the "cutoff point" is after the first neighbor.[106]

The last piece of structure that we need is a probability function that is defined on the set of states. In the ferromagnetic case, the natural choice is the canonical distribution.[107] In the case of the gas–liquid transition, the number of degrees of freedom might change, so the chosen ensemble is the grand canonical one.

Now we are in a position to state the abstract form of an argument that establishes the existence of phase transitions in a two-dimensional square Ising lattice. When the argument is applied to ferromagnets we obtain an explanation of "spontaneous magnetism." As before, we will begin by discussing the argument abstractly and then apply it to ferromagnetic phenomena.

---

[105] This anthropocentric language should be understood, as usual, in terms of the variational principle that systems attempt to reduce their level of potential energy to the minimum. When σ_i and σ_j are the same, $\sigma_i\sigma_j = (-1)(-1) = 1$ or $\sigma_i\sigma_j = 1 \times 1 = 1$. In both cases, when $J > 0$, $-J\sigma_i\sigma_j < 0$. When σ_i and σ_j are different, $-J\sigma_i\sigma_j > 0$.

[106] The main results, however, remain valid even if less restrictive assumptions are introduced.

[107] The reason for choosing the canonical distribution is that it is the only distribution that is at the same time translation-invariant and whenever it is restricted to a finite region it coincides with the counting measure. Note that translation-invariance is completely analogous to stationarity.

The first step in the argument is to associate a macroscopic property of interest with a "global" lattice property. For example, we associate with magnetization a "global polarization" property. Magnetization is defined as $N^+ - N^- / N^+ + N^-$ (N^+ is the number of positive spins and N^- is the number of negative spins[108]). Spontaneous magnetization occurs when there is a nonvanishing magnetization in the absence of an external magnetic field (which means that most of the spins will align in a parallel way along the z-direction). Next we specify the set of "critical conditions" that are necessary for the property of interest to arise. For example, spontaneous magnetization of ferromagnets happens at the Curie temperature (for iron, the critical temperature is 1043 K). The set of "critical condition" constraints are those states that are consistent with this temperature. The main part of the argument consists of a proof that when the "critical condition" constraints reached the global property will emerge with an overwhelming probability. In our example, we have to establish the fact that spontaneous magnetization is overwhelmingly likely at the Curie temperature.

The proof of the existence of phase transitions is rather technical, and we shall not attempt to reproduce it in its entirety. Nevertheless, some aspects of the proof are quite important for understanding the status of the Ising–Lenz models. Therefore, we shall proceed with a set of brief remarks that pertain to the most noteworthy aspects of the argument. The most important issue that emerges is that the proofs of the existence of phase transitions do not proceed purely from first principles. They rely on the assumption that thermodynamic properties are "bulk" properties, that is, they are insensitive to the precise character of the boundary conditions.[109] (We shall refer to the above premise as premise P). If a proof from first principles existed, we could have regarded it as a reduction. It would have demonstrated that a well-known fact from thermodynamics can be derived from mechanical principles.[110] The existing proofs, on the other hand, cannot be

---

[108] In "disordered states," that is, those states that do not exhibit magnetic properties, N^+ is going to be roughly equal to N^-, in which case the expression $N^+ - N^-/N^+ + N^-$ will be equal to zero.

[109] To understand why this premise is true, observe that when we place an ice tray in the freezer we do not expect the shape of the ice tray to influence the behavior of the water. (For example, we do not expect the shape of the container to influence the freezing temperature.)

[110] The reader should note, though, that the notion of reduction needs to be clarified because, in our cases, as in most of statistical mechanics, the arguments invariably contain a statistical element.

regarded as reductions because they rely on a premise that cannot be derived from mechanical principles. In fact, as we shall see, the next missing premise is not only independent of the mechanical framework, it is also dissonant with it. The motion of the individual molecules is extremely sensitive to the shape of the boundaries. Therefore, an observer who is acquainted only with the microlevel would never even entertain P, let alone use it as a premises in an explanation. Therefore, the picture that Hempel,[111] for example, presented cannot be made to apply to our case; the very notion of explanation that is operative in statistical mechanics is different than Hempel's. Rather than deriving the explanation from our existing microtheory, we derive it from a new theory that is itself dissonant with the microtheory. Let us now say a few words about an actual argument that establishes the existence of phase transitions. This argument was developed by Peierls in 1936.[112]

Let us return to the lattice model. As you may recall, each site in the lattice is either "+" or "−" (spin up or spin down). Therefore, a pair of neighboring sites may be in one of the following four positions: ++, −−, +−, −+. Let us suppose that ++ and −− have −J as their interaction energy and +− and −+ have +J as theirs. As we remarked, if J is positive, like spins "prefer to stay together" and the model is ferromagnetic; if J is negative, the resulting model is antiferromagnetic. In antiferromagnetic models, the spins are arranged in an alternating pattern. The formation of such a pattern is used in a variety of different contexts to explain the formation of crystalline structures. (Crystals are molecular formations that obey certain spatial patterns.) Let σ be a variable that takes the values +1 and −1. (Observe that $\sigma^2 = 1$ regardless of the value of σ). A more explicit formulation of the "law" that determines the development of the configuration of the spins is as follows:

$$-J\sum_{i,j}\sigma_i\sigma_j - mh\sum_i\sigma_i,$$

(E)

where h is the external magnetic field, m is the magnetic moment of the spin and, as before, where the first sum is taken over nearest-neighbor pairs. Now suppose that we know T, the temperature of the system. Because the probabilities are calculated from the canonical distribution, we are guaranteed that at a given temperature there will be

---

[111] See his *Aspects of Scientific Explanation*, Free Press (New York, 1965).
[112] See Peierls, "On Ising's Model of Ferromagnetism," *Proc. Cam. Phil. Soc.* 32 (1936), pp. 477–481.

one energy level that is vastly probable. Let E be this energy level. Let us further assume that there is no external magnetic field that is operating on the system.[113] In such a case, it can be demonstrated that the only terms that contribute to E are sites that have nearest neighbors with an opposite spin. Therefore, E is determined by the lengths of the borders that separate a homogeneous group of spins from its environment. (Upon reflection, it should be obvious that the determination of the borders fixes the state of the system uniquely, or, rather, up to a simultaneous flip of all of the spins.) Let us make one more simplifying assumption, namely, that we may ignore borders that end on the boundary of the lattice. (The reason for the assumption is, again, that we want to ignore the effects of the boundaries.) With this assumption we can formulate the question concerning the existence of phase transitions rather easily. We ask whether there is a critical temperature T_c such that, in the thermodynamic limit for $T < T_c$, the total area occupied by enclaves containing "deviant" spins will be a "negligible quantity." The first demonstration of this statement was given by Peierls. His argument was crude, but it applies in a surprising variety of cases. His strategy was to look for an upper bound of the total area of the deviant spins and to show that for sufficiently low temperatures the expression of this upper bound converges. The argument was refined in the 1940s by Onsager, who gave a complete solution to the question of the existence of phase transitions in the two-dimensional ferromagnetic case. In the 1960s, Lee and Yang used the Ising–Lenz model to demonstrate the existence of phase transitions between the liquid and the solid phases.[114] An important point that should be emphasized is that our intuitive notion that phase transitions involve a certain measure of discontinuity is confirmed by the theory of phase transitions. When phase transitions take place, the first derivative of the function that correlates magnetization with density has discontinuous points. Below a certain temperature, the system will be in one phase;

---

[113] Contrary to what it may seem, these assumptions do not limit the generality of the argument.

[114] In this case, as well as in the case of the gas–liquid transition, an occupied site receives the value +1 and an empty site receives 0. Observe that in this case, too, we may use the same Hamiltonian. When the phase transition occurs, most of the microscopic cells become occupied. This fact is expressed by again choosing $J > 0$. In such a case, $-J\sigma_i\sigma_j < 0$ only when σ_i and σ_j are both occupied. In all of the other cases, $-J\sigma_i\sigma_j = 0$.

above it, it will be in another. The precise dynamics of the transformation, although continuous, is not completely predictable. The existence of such a subtle connection between the empirical phenomenon and the mathematical model is generally considered rather remarkable.

Another noteworthy aspect of the Ising–Lenz argument is that it entails a solution to the following puzzle of symmetry breaking. Spontaneous magnetization can be realized in two different ways, one in which the majority of the spins are oriented "down" and the other where most of the spins are "up." The two possibilities are completely symmetric in the absence of an external magnetic field. So, how does the system "choose" between the two options? The answer is the following. Suppose that we observe an Ising–Lenz system that is under the influence of an external magnetic field. We then let the field approach zero gradually. It can be demonstrated that, even after a long "relaxation time," the system "remembers" its initial state, that is, it will exhibit a spontaneous magnetization in the direction of the now-defunct external magnetic field. This sensitivity to "initial conditions" is quite puzzling. First, we encounter a system whose future is not "asymptotically independent" from its initial state. Hence, it is one of the cases in statistical mechanics in which one encounters "nonergodic" systems. Next we should note the fact that an arbitrarily small difference in the initial external magnetic field might lead to an entirely different limit behavior. (This observation is the basis of a proof of a fact that we mentioned earlier, that is, the function that connects magnetic behavior to potential has a noncontinuous first derivative at the point where the external field equals zero.) Next observe that, although there are two possible magnetic states that the system might assume, once the system assumes a magnetic state it will continue to do so even in the face of perturbations. It is quite easy to see that, for Ising systems of two dimensions or more that are in a sufficiently low temperature, the probability of a "spontaneous" energy fluctuation that is sufficient to destabilize the magnetic state is very small. Therefore, we may say that, in the absence of an external field, the system "chooses the direction of its spontaneous magnetization randomly and sticks to it."

Let us now sum up the case of the Ising–Lenz theory. When Lenz introduced his theory of ferromagnetism, he said that as a result of his theory "the magnetic properties of ferromagnets would then be

127

explained in terms of nonmagnetic forces."[115] In a sense, his ideas were proven right, and, consequently, one may regard the Ising–Lenz theory as a reduction. However, the argument that establishes the existence of phase transition depends on a premise that is not only independent of the mechanical background but is dissonant with it. The premise is that macroscopic properties are "bulk" properties that are not sensitive to the boundaries. The inclusion of this premise disqualifies the theory from being a reduction proper because the explanation does not follow from a set of purely mechanical principles. Therefore, the Ising–Lenz theory is essentially an inductive consistency argument. As we mentioned earlier, the theory is extremely useful, and to this day it is used to derive information concerning the behavior of systems near their critical points. The fact that the theory does not proceed from first principles does not render it any less useful. However, a further question concerning the theory emerges: How did it come about that the community of physicists accepted a theory with this type of shortcoming? What can we learn from this example about the relations between theoretical depth and explanatory power? To answer these questions, let us begin with a short digression on the history of the Ising–Lenz theory.[116]

When Lenz originally proposed his theory of ferromagnetism in 1920, he thought that his lattice model furnished a fairly realistic description of ferromagnets. His assumption was that the dipolar atoms rotate around a fixed rest position in the crystal. The physical basis for the assumption has to do with the peculiarities of the "old quantum theory." Ising and others believed that the dipolar atoms have only a limited number of possible orientations. Therefore, it was natural for him to build a model that takes into account only the distribution of these orientations among the different sites. Lenz assigned his student, Ising, the task of proving the existence of phase transitions in the simple one-dimensional case, and Ising proved that phase transitions cannot arise in this case. But if it were possible, and the results were different, the theory would have been perceived at the time as a fairly realistic account of ferromagnetism. A few years after Ising published his negative results, Heisenberg published his own account of ferromagnetism, which did not exclude the possibility of rotations.[117]

---

[115] S. G. Brush, *Statistical Physics and the Atomic Theory of Matter*, p. 241.
[116] For further reading, see ibid, pp. 233–259.
[117] See Heisenberg, "Zur Theorie Ferromagnetismus," *Z. fur Pyz.* 49 (1928).

128

Heisenberg's theory involved a more complex interaction between spins. After that, the idea of using the lattice model to model ferromagnetism remained dormant for a while.

In the meantime, an analog of the lattice gas theory was investigated as part of the research into order–disorder transformation in alloys. (In particular, Bethe invoked the nearest-neighbor interaction to improve the Bragg–William theory of order–disorder transformation.) Peierls, who followed Bethe's research, understood that there was an equivalence between Bethe's work and the work on Ising–Lenz models. This realization led him back to the lattice gas model of ferromagnetism and to develop the argument for the existence of phase transitions in the two-dimensional case. At this point, though, the status of the Ising–Lenz model had changed. Peierls, who was acquainted with Heisenberg's theory of ferromagnetism, did not consider the Ising–Lenz model a realistic theory of ferromagnetism. In fact, he explicitly stated that the research on the Ising–Lenz model had only a "mathematical interest."[118]

The major breakthrough, though, occurred in 1944, when Lars Onsager published a complete solution of the problem of the two-dimensional lattice model.[119] This solution was remarkable because, for the first time, it was possible not only to predict the existence of phase transitions, but also to calculate the rate in which the transition occurred. Nevertheless, even after the publication of Onsager's results, physicists did not take the Ising–Lenz model very seriously. According to Domb, this attitude was derived from the fact that "the two-dimensional model was conceived as highly artificial and did not represent exactly any known physical system."[120]

Another point to be made is that the Ising–Lenz theory was not the only theory of critical phenomena. From the 1940s to the 1960s, there was an active attempt to give more "realistic" explanations of phase transitions. In particular, there was an attempt by Meyer and his students to provide a formulation of the theory of phase transitions in terms of continuum mechanics. These attempts were quite successful, and, at the time there was no reason to prefer the Ising–Lenz theory to Meyer's theory. However, in 1961 a mathematical technique was

[118] Ibid, p. 477.
[119] See L. Onsager, "Crystal Statistics I. A Two-Dimensional Model with an Order–Disorder Transition", *Pys. Rev. Ser. 2* 65 (1944).
[120] Quoted in Brush, ibid, p. 246.

developed that enabled physicists to obtain further information about the singularities at the critical point from some of the known terms in the series expansion of the Ising–Lenz model. This technique was used to calculate a much better approximation of the critical exponent, which allowed a better calculation of the intensity of the spontaneous magnetization in the neighborhood of the critical point. Consequently, by 1965 there was no doubt that the Ising–Lenz theory was far more successful in treating critical phenomena than the more "realistic" continuum theories. And for those who wondered how it is possible that a strangely impoverished theory can be so informative, there was the beginning of the explanation. It was conjectured that "the closeness of the values provides support for the view that the dimensionality and the statistics are the main factors that determine critical behavior, the details of the interactions (which are grossly oversimplified in the lattice model) being of secondary importance."[121]

Nevertheless, even after these developments, some physicists felt that something was amiss. Uhlenbeck gave a voice to these sentiments when he lamented the fact that numerical calculations do not provide a "general point of view." He also remarked that, "only in the neighborhood of the critical point do the particles remember the work of Onsager."[122] Consequently, Uhlenbeck hoped that a "reconciliation will be reached" between the Ising–Lenz theory and the "classical theories" that might be traced to the work of van der Waals. According to Domb, Uhlenbeck's hope was fulfilled, at least to a considerable extent, by the results of his own research and that of others.[123]

What, then, can we learn from this historical episode? The following is a list of conclusions that we have obtained by reflection on the history of the Ising–Lenz theory.

1. The most obvious point to be made is that when a theory offers a solution to a well-known and disturbing puzzle, physicists will lend to it a sympathetic and attentive ear. This, of course, does not mean that the shortcomings of the theory, if there are any, will be ignored. We only suggest that the disagreeable aspects of the theory will be weighed against its potential benefits. Therefore, the bigger the

---

[121] See Brush, ibid, p. 253.
[122] Ibid, p. 256.
[123] Ibid.

benefits (the more troublesome the puzzle), the more sympathetic the ear.

2. We do not judge theories in isolation but compare them to their "competitors." In particular, if a theory is clearly superior to its alternatives, its disagreeable aspects will be treated with relative lenience. If there are rival theories that are almost as good while also being problem-free, the disadvantages of the theory will weigh more heavily against it.

3. The next point is also rather straightforward. We assumed all along that all of the theories we are discussing agree reasonably well with the observed data. However, if the theory entails extremely accurate predictions, and if these predictions are relatively robust, the theoretical assumptions of the theory will be taken more seriously, even if they are somewhat problematic in their own right.

4. Finally, a theory that exploits its mathematical structure in an effective and novel way is likely to be taken seriously by the scientific community. The assumption in this case is that the theorist might have stumbled on a new mathematical discovery that is yet to be given a precise formulation.[124] Unfortunately, this important point is not discussed much by philosophers of science.

Note that in each of the four remarks we introduced a factor that is not directly related to the relation between the theory and the systems that it attempts to explain. Nevertheless, we maintain that these factors are relevant for the determination of the conditions under which the theory will be accepted by the community of physicists. Therefore, we conclude that the notion of explanation cannot be a purely "semantic" notion. To be sure, issues such as completeness and naturalness, which invoke the relation between the theory and the systems that it describes, do seem relevant to the physicist. They determine how attractive the theory is prima facie. But the ultimate verdict, whether the theory will be accepted or not, will be a function of other factors as well (for example, by the four factors outlined in this section). The case of the Ising–Lenz theory demonstrates how effective these factors can be in overriding the damaging effect of an apparent lack of depth.

---

[124] Mathematical discoveries often announce themselves first through their applications to physics. When this happens, the physical theories themselves at first seem awkward and devoid of direct physical reality. These peculiarities often cease to be troublesome once the new mathematical theory receives a proper formulation.

131

2.11. BEYOND INDUCTIVE CONSISTENCY:
THE RESULTS OF SINAI

Before we describe the results of Sinai, let us first recall the basic features of the so-called Boltzmann–Gibbs billiard-ball table model of an ideal gas. The model represents the gas molecules as hard spheres contained in a rigid box. The spheres collide with each other in a perfectly elastic way, obeying the law that the angle of reflection is the same as the angle of impact. Suppose, for the sake of simplicity, that we have only two spheres and that one of them is fixed. In other words, imagine that only one of the spheres is actually moving. If the "obstacle" (the fixed sphere) is convex, small differences in the angle of impact will yield exponential differences in the angle of reflection. This means that the distances between nearby trajectories will be amplified upon reflection, at least for a while, until only one of them hits the obstacle while the other escapes, and, from that point on, the subsequent evolution of the two trajectories will be totally independent. The dynamics of the balls in the billiard model has an interesting property – the motion along each of the trajectories will be constantly locally unstable. This property is a verifiable local property of each of the trajectories. Sinai proved that such a motion is ergodic, which implies that the limiting average behavior of almost every trajectory is the same, relative to the natural Lebesgue measure. In other words, the average limiting behavior will be almost always the same, independent of the initial direction of the departure from equilibrium. This means that the probabilities we should assign to the events that the sphere will pass through certain regions on the table should be relatively insensitive to additional information about the initial condition of the system. Notice that, even if we were omniscient observers, we would appreciate the fact that the motion of the trajectories is unstable. Therefore, the fact that the probabilities are related to a local property of the individual trajectories supplies them with an empirical content that remains meaningful even if the probabilities are no longer needed for the purposes of prediction.[125]

Instability also explains why ignorance with respect to the exact (or even the approximate) state of an unstable system is to be expected.

---

[125] The point echoes some of von Smulochowsky's remarks. See Smulochowsky, "Uber de Begriff des Zufalls und den Ursprung der Wahrscheinlichkeitsgesetze in der Physik," *Die Naturwissenschaften* 6.

The existence of singularities (e.g., because of the collisions) makes it difficult to calculate the state of such systems, and the nonlinearity of the motion makes it difficult to employ approximation techniques. So instability furnishes us with an "objective" reason for the need to use probabilistic reasoning.

Poincaré, who was one of the first thinkers to appreciate the importance of instability, demonstrated the emergence of stochastic behavior as a result of instability using the following example. Suppose that we are trying to determine whether a ball that falls down a bumpy mountain will end up in area A, located at the foot of the mountain. Assuming that the descent from the mountain is free, the only missing information concerns the initial direction of the ball, that is, the initial direction of its departure from its unstable equilibrium. If we know the precise direction, it is possible, in principle, to determine whether the ball will end up in A or not. In this case, though, approximations are completely useless. Therefore, we may use the relative size of A as an estimate for the probability that the ball will end up in A. In other words, even if we had some additional information, we could still do no better than using the relative size of A as the basis for fixing our degrees of belief.

Sinai's results point to the fact that one may attribute to statistical mechanics a role that is significantly more ambitious than merely affecting an inductive constancy proof. His work demonstrates that one may aspire to derive the thermodynamic behavior from purely mechanical principles rather than merely proving that such behavior is consistent with the mechanical principles. Moreover, Sinai's results show that the emergence of "chaos" is not an artifact of the use of coarse grain descriptions[126] but instead is connected to the details of the fine grain details.[127] In the following chapters we shall continue this line of reasoning; we shall investigate the dynamics of n-particle systems and use our results to illuminate thermodynamic phenomena.

[126] The fact that the motion is represented as an action in a space of negative curvature means that each of the trajectories changes very rapidly. It is this property that is responsible for the ergodic behavior.
[127] This aspect of Sinai's work does not carry over very well to quantum statistical mechanics.

3

The Haar Measure

3.1. GROUP-INVARIANT MEASURES

In 1933, Alfred Haar wrote a paper, entitled "The Concept of Measure in the Theory of Continuous Groups,"[1] whose main objective was to construct a measure invariant to the action of a topological group. The importance of the paper must have been noticed immediately, and further studies followed shortly on its heels. Von Neumann proved the uniqueness of this measure in 1934.[2] In 1937, Stefen Banach wrote an appendix about the Haar measure for a textbook about integration theory,[3] and in 1940 both Weil[4] and Cartan[5] offered new uniqueness proofs. What is interesting and important about Haar's construction is that it leads (in most important cases) to a measure that is different from the Lebesgue measure by only a factor of proportionality. However, because the Haar measure is more general and abstract, it can illuminate and even justify the choice of the Lebesgue measure as the natural measure. In particular, the construction makes it easier to explain why we use the Lebesgue measure as a probability measure when it is defined as a volume measure.

The Haar measure is hardly unknown to mathematicians. In fact, for many mathematicians the Haar construction constitutes the foundation of their views on probabilities. It contains existence and uniqueness proofs in a fairly general setting, and it coincides with our

---

[1] See A. Haar, "Der Massbegriff in der Theorie der Kontinuierlichen Gruppen," *Ann. Math.* 34 (1993), pp. 147–169.
[2] See von Neumann, "Zum Haarschen Mass in Topologischen Gruppen," *Comp. Math.* 1 (1934), pp. 106–114, and "The Uniqueness of Haar Measure," *Mat. Sbor.* 1 (1936), pp. 721–734.
[3] See, S. Banach "On Haar Measure" in Saks' *Theory of the Integral.*
[4] See Weil, *L'Integration dans les groupes topologiques et ses applications*, Herman & Cie (Paris).
[5] See Cartan," Sur la mesure de Haar," *C.R. Acad. Sci. Paris* 211 (1940), pp. 759–762.

pretheoretical notions of how probability measures ought to behave. However, the importance of the Haar measure is not due solely to its mathematical character. Because of its highly abstract formulation, the Haar measure lends itself to a variety of different interpretations, which makes it significant for a range of different approaches to the foundations of probability.

To begin with, the Haar measure marks the most unproblematic case in which probabilities may be interpreted as the limits of relative frequencies. In the case of the Haar measure, there are no further probabilistic concepts lurking in the background, so there is no danger of circularity when probabilities are defined as relative frequencies. It is well known that many eminent physicists, including Boltzmann and Einstein, held the frequentist account of probabilities. These physicists hoped that the probabilities that appear in physics could be identified with the appropriate relative frequencies. This hope formed the basis for what is often referred to as *Boltzmann's ergodic hypothesis*[6]: If a physical system passes through all of the points in its phase space as it evolves in time, it is possible to imagine that the probabilities that it will assume certain states may be identified with the frequencies with which it occupies these states. The Haar construction transforms this intuition into a proof that this is indeed true in some limited cases. Unfortunately, once we follow the construction, we see clearly the impossibility of extending it to a more significant class of physical systems.

This explains the significance of the Haar measure to frequentists. However, the construction is extremely significant for the logicists as well. Recall that, according to the logicist conception of probability, to fix our probabilities we first must find a relation of equivalence or congruence on the space of events relative to which the probabilities remain constant. The hope of the logicists (or the symmetricists, as they are sometimes called) is that such a relation could determine the probabilities uniquely. And, once more, the Haar construction shows the extent to which the hopes of the logicists may be fulfilled: Some relations of congruence determine the probabilities uniquely. It will become equally clear, though, that the logicist faces a serious dilemma. If the relation of congruence on the space of events is based on a *phys-*

---

[6] In "Boltzmann's Ergodic Hypothesis," in *Archive for History of Exact Science*, von Plato argued rather persuasively that, in fact, Boltzmann never espoused this version of the ergodic hypothesis.

ical concept of congruence, the logicist's conception suffers from the same problems as the frequentist's conception. If, on the other hand, this relation is based on a subjective judgment, then logicism should be taken as a peculiar formulation of personalism. In this sense logicism is an incomplete view – the relation of symmetry itself must be explicated in either physicalist or personalist terms.

Another, less familiar use of the Haar measure is as a method of eliminating probabilities. In this context, the concept of elimination is not the subjectivist one, where probabilistic statements are analyzed as propositional attitudes. What is meant by eliminating probabilities in this context is that, if A is a statement containing probabilistic concepts, it is eliminated when it is replaced by an equivalent statement B that does not contain any probabilistic concepts. This concept of elimination belongs to the tradition of Hilbert, Quine, and Field. The comparison to Hilbert seems especially interesting. As the reader may recall, Hilbert's program of eliminating infinitary concepts and proofs from mathematics came as a response to a host of problems to which the concept of infinity gave rise. First, there are problems concerning our ability to comprehend infinite totalities and proofs. It seems that every epistemological account of infinitistic mathematics involves ad hoc assumptions. Another issue concerns the applicability of infinitary concepts to physics. Even if we had a satisfactory account of the ability to comprehend infinite totalities as abstract objects, it would still be necessary to ask how we could obtain measurable parameters from infinitary concepts that would allow us to apply these concepts in physics. And finally there is the issue of justifying infinitary inferences such as indirect existence proofs.

The concept of probability raises similar issues. In this context, too, we must explain how we come to know about probabilities, how can we apply them in a scientific context, and how we can justify probabilistic inferences. Obviously, more needs to be said to prove that the similarities between the two cases point to something deeper than a superficial analogy. However, in this case as well it is tempting to ask whether we really need probabilities. That is, is it possible to eliminate probabilites by proving that every sentence or proof using probabilistic concepts can be replaced by equivalent sentences and proofs that do not contain probabilistic concepts? As we shall see, Haar's construction can be seen in precisely this light. In particular, his construction will give rise to an idealized instrumentalist conception of probability. But, here, too, arises the question of the application of this method to physically significant cases. Unfortunately, there are not many significant applications of this remarkable result.

3.2. CONSTRUCTION OF THE HAAR MEASURE

In this section we shall present the main idea of Haar's original paper. The presentation will follow Banach's appendix in Saks's "Theory of the Integral".[7] For the actual proofs, the reader may consult Nachbin[8] or Halmos.[9]

We shall start with a "space of events" $\langle S, \mathsf{B} \rangle$, where S is a set and B is a σ-algebra of subsets of S. If the notion of an open set is given for S, B should be a Borel σ-algebra (i.e., the σ-algebra generated by the open sets). $\langle S, \mathsf{B} \rangle$ is also assumed to be metric and locally compact.[10] In addition, we require that there be a relation of congruence "$\cong$" defined on $\langle S, \mathsf{B} \rangle$. "$\cong$" is a relation between members of B satisfying the following axioms (all of the sets mentioned below are members of B):

1. "$\cong$" is reflexive, symmetric, and transitive.
2. If A is a compact, open set and $A \cong \mathsf{B}$, then B is open and compact.
3. If $A \cong \mathsf{B}$ and $\{A_i\}_{i=1}^{n}$ is a sequence of open sets s.t. $A \subset \cup_{i=1}^{n} A_i$, then there exists a sequence $\{\mathsf{B}_i\}_{i=1}^{n}$ such that $\mathsf{B} \subset \cup_{i=1}^{n} \mathsf{B}_i$ and for all i, $A_i \cong \mathsf{B}_i$.
4. For every compact open set A the class of sets congruent to A covers S.
5. If $\{S_i\}$ is a sequence of compact concentric spheres with radii tending to zero and $\{G_i\}$ is a sequence of sets s.t. $G_i \cong S_i$, and if $\{a_n\} \subset G_n$, $\{b_n\} \subset S_n$, $\lim a_n = a$, and $\lim b_n = b$, then $a = b$.

It can be shown easily that if A and B are compact, open sets, there is a finite sequence of sets congruent to A whose union covers B. Therefore, we can define $h(A, \mathsf{B})$ as the least number of sets congruent to A whose union covers B. The number $h(A, \mathsf{B})$ is a rough approximation of the relative sizes of A and B. It is rough because these sets cover B with a residue. Let us now fix a "large," compact open set G. The quantity

---

[7] See ibid.
[8] See Nachbin *The Haar Integral,* Van Nostrand (Princeton, NJ, 1967).
[9] See P. R. Halmos, *Measure Theory,* Van Nostrand (Princeton, NJ, 1950).
[10] Where compact spaces are concerned, every collection of sets that cover the space contains a finite subcollection that also covers it. This requirement is satisfied by the representations of the phase spaces of bounded systems. In locally compact spaces, every set is contained in a compact set.

$$h^*(B) = h(A,B)/h(A,G)$$

will be a rough approximation of the size of B. This approximation will become more and more accurate as A gets smaller and smaller. Let us now take, instead of A, a sequence S_n of concentric spheres located in G with radii tending toward 0. The quantity

$$m(B) = \lim_{n \to \infty} h(S_n, B)/h(S_n, G)$$

is that which interests us. We extend it to all of the members of B, and then we prove that it is a measure, that is, that it is monotonic with respect to inclusion, sigma-additive, and receives the maximal value of 1 for G.

So far, the choice of G was the only arbitrary step in the construction. However, it can be shown that the choice of a different "large" set yields only a difference in the choice of a factor of proportionality. As we have mentioned, it was proven that in a compact metric space the above-mentioned concept of congruence yields an essentially unique concept of measure.

3.3. INTERPRETATION AND APPLICATION OF THE NOTION OF CONGRUENCE

A relation of congruence between A and B can be interpreted in one of the following two ways:

1. A and B have a certain property in common, that is, they have the same size or area.
2. It is possible for A and B to "evolve" into each other.

The first interpretation is appropriate for "simple" objects, such as squares or line segments, but when we want to extend the notion of size to more complex sets, the second interpretation is more natural. We select a class of transformations that preserves the area of the simple objects, and we define congruent sets as those that can be transformed into each other by some transformation from this class.

Therefore, instead of starting with the space $\langle S, B \rangle$ on which the relation "$\cong$" is defined, we can start with a group M of one-to-one and continuous transformations $T: S \to S$ such that, for all $A \in B$ and for all $T \in M$, $T(A) \in B$ iff $A \in B$. Being a group, M has the following properties:

138

P₁. $T \in M$ implies that $T^{-1} \in M$.

P₂. If $T_1, T_2 \in M$, then $T_1 T_2 \in M$, where $T_1 T_2(x) = T_1 (T_2(x))$.

In addition, we suppose that

P₃. For all $a,b \in S$ there is $T \in M$ such that $T(a) = b$.

P₄. If $\{a_n\}$ and $\{b_n\}$ are converging series, such that $a = \lim a_n = \lim b_n = b$ and $T_n(a_n)$ and $T_n(b_n)$ are converging for all n, then

$$T_n(a) = T_n(b).$$

Therefore, given $A, C \in B$, we may define $A \cong C$ iff, for some $T \in M$, $T(A) = C$. It can be proven that this definition fulfills all of the conditions that we set forth for the relation "$\cong$".

An especially interesting case is the one in which M is created by the iterates of a single transformation T. An example of this kind may be obtained from the rotations of a circle. If T is thought of as an infinitesimal rotation of the circle, it can be shown that a repetitive iteration of T will generate the class of all of the rotations of the circle, a class that clearly obeys the above requirements $P_1 - P_4$.

It is interesting that this case applies to the mathematical description of the evolution of a system with one degree of freedom. Take the following example.[11] Let b be a ball that moves with a constant velocity along a frictionless rail. It is not hard to see that the evolution of such a system can be given a topological group representation. As the space S, we shall take the configuration space or the space of the possible positions of b. B will include the Lebesgue measurable sets and $T_\epsilon : S \to S$ will be the infinitesimal "motion" along S. The Haar construction guarantees that there is a measure m on $\langle S, B \rangle$ that is invariant with respect to T_ϵ. It is easy to see that m will be essentially identical with the circular Lebesgue measure on the circle S.

Remarks

1. *Given a set* A $\in$ B, m(A) *can be measured by the relative amount of time that* b *spends in* A. *That is, if* t(A) *is the time spent in* A, *then* m(A) *is the ratio between* t(A) *and* t(S).

2. *It is possible to approximate* m(A) *by calculating the ratio between* t(A) *and* t(G), *where* G *is a large compact set.*

---

[11] Gibbs talks about similar examples in *Elementary Principles in Statistical Mechanics*, Yale Univ. Press (New Haven, CT, 1948), pp. 142–143. Von Plato refers to a discussion by Einstein that involves similar examples as well. See ibid.

3. *If* A *is not a connected set, the ball* b *will enter and exit the different parts of* A. *In such a case, too, it is clear that* m(A) *measures the relative frequency of the stay in* A.

4. *If* t(S), *as measured by an ordinary clock, is not finite,* m(A) *must be calculated by a limit operation, that is, relative to a series* $G_i \to S$, *of compact sets such that* t(G_i) *is finite. In such a case, the interpretation of the magnitude* m(A) *as a measure depends on the assumption that the convergence to the limit is uniform.*

The example that we are dealing with is extremely superficial from a physical point of view. As we shall see shortly, it is possible to go beyond this contrived example, but unfortunately not very far. The crux of the matter is requirement P_3. If T_ε is to represent the equations of motion, P_3 requires that the system must realize all of its possible states as it evolves; if it does not, it cannot be represented as a topological group. This requirement, however, cannot be realized when S has a dimension greater than one. In higher dimensions, the evolution of the system is represented by a curve l in a space of a larger dimension, and for reasons of dimensionality l cannot fill such a space.[12] This conclusion is especially obvious when S is the $6n$-dimensional phase space of a system with a large number of degrees of freedom.

When the space S has a dimension larger than one, there will be many trajectories that the system could take; relative to these different evolutions, different time averages are to be expected. As we have seen, the celebrated ergodic theorems state the conditions under which, for a "large" set of initial conditions, the respective time averages of stay in the different sets will be identical to the measure

---

[12] Peano constructed a space-filling curve, but the correspondence is not one to one. See Peano, "A Space Filling Curve" in *Selected Works of Giuseppe Peano* (Univ. of Toronto Press (Canada, 1973), pp. 143–149. "Mr Cantor," he writes, "(...) has shown that a 1–1 correspondence between the points of the line and those of a surface can be established. But Mr. Netto [*J. Reine Angew. Math.* 86 (1879), pp. 263–268] and others have shown that such a correspondence is necessarily discontinuous. [See also G. Loria, "La definizione dello spazio ad n dimenzioni . . . secondo le ricerche di G. Cantor," *G. di Mathematiche* (1877).] My note shows that single valuedness and continuity from one side, i.e., to the points of a line, can be made to correspond to the points of a surface in such a fashion that the image of the line is the entire surface, and the point on the surface is a continuous function of the point of the line. But this correspondence is not one-to-one." See ibid, p. 147. Note that, for our purposes, both the continuity and the 1–1 correspondence are essential. Therefore, Peano's construction does not have a significant physical interpretation.

of the set relative to a "natural" measure m (e.g., the Lebesgue measure).

It is worth noting that the Haar measure is conceptually different from the Lebesgue measure when the latter is construed from inner and outer measures of sets.[13] Hence it is possible to regard the identity between the two as an especially trivial ergodic theorem: The time averages (the Haar measure) are identical in value to the space measure (the Lebesgue measure). However, if the usual ergodic theorems state that the identity is true almost-always, in our case the identity *always* obtains. Therefore, the replaceability of the Lebesgue measure with the Haar measure has foundational significance for the frequentists.

As we remarked earlier, Boltzmann, motivated by frequentism, introduced into SM the ergodic hypothesis, an assumption that is essentially equivalent to P_3.[14] The assumption states that systems with many particles will pass through all of their possible states as they evolve. As we shall see shortly, it is fairly evident that in most cases physical systems do not satisfy the ergodic hypothesis. Because of this, Boltzmann and other frequentists had to find a more acceptable formulation of their idea. As we shall see, these attempts are all rather problematic from a foundational point of view.[15]

A slightly more realistic physical system with one degree of freedom is a pendulum that moves in a frictionless space. Because the pendulum moves 180° and then returns, the velocity vector changes its sign on the way back. Therefore, the configuration space of the pendulum (i.e., the space of possible positions) can be plotted as a circle C. The Haar construction guarantees that there is a motion-invariant measure m. In this case, however, m is not identical to the Lebesgue measure and so the measure $m(A)$ is not proportional to the time of stay $t(A)$. The reason for that is that the pendulum spends more time in areas near the extreme points a and b than near the center c. However, the Lebesgue measure $M(A)$ of a set A can be represented as $m(A)f(A)$, where f is a function that depends on the velocity of the pendulum inside A. In this case, then, the Haar construction must be supplemented by "weights" to yield the Lebesgue

---

[13] For the definition of a Lebesgue measure in measure-theoretic terms, see Halmos, ibid.

[14] For the history of the formulation of the hypothesis, see von Plato, ibid.

[15] We discussed many points that are related to this issue in the Chapter 2.

measure.[16] Because the weights are themselves measurable, this case too is significant from a foundational point of view.

Let us now summarize what we take to be the significance of the Haar construction for the frequentists. The frequentist's conception of probabilities is first generalized by the concept of time averages. (This issue was discussed in Chapter 2).[17] Next we use the Haar construction to give an explicit definition of the time averages. Then we interpret the Haar measure as a time average in systems with one degree of freedom. Finally, we observe that the Lebesgue measure is essentially identical to the Haar measure. We arrive in this way at a complete justification of the choice of Lebesgue measure. The Lebesgue measure is, essentially, the only probability function on $\langle S,B \rangle$ that can be interpreted as a measure of relative frequencies. What remains to be seen is whether the case of the system with one degree of freedom is representative of the larger family of cases wherein probability arises.

3.4. WEAK CONGRUENCE RELATIONS

As the reader may recall, all that is required to carry out the Haar construction is an abstract relation of congruence on the state space. Consequently, if we could justify the principle that all of the congruent sets should be equiprobable, we would need no further justification for choosing the Haar measure to guide our probabilistic reasoning. So, for the logicist, whose probabilities reflect a reliance on a conception of symmetry or congruence, or a principle of indifference between certain events, the Haar construction is extremely useful. Given a relation of indifference or symmetry (formulated as a relation of congruence), the Haar construction constitutes a sufficient condition for the relation to yield a unique probability function, namely, if the relation satisfies conditions 1–5, there is an essentially unique assignment of probabilities relative to which symmetric events will be equiprobable.[18]

---

[16] If A is a segment $[a,b]$, $M(A) = m(A) \frac{1}{2}(v_a - v_b)$, where v_a, v_b are the velocities at the points a,b, respectively. The extension of this idea to other measurable sets is straightforward.

[17] The generalization is analogous to the construction of integrals from step functions.

[18] Jaynes wrote about the relevance of invariance principles to logicism in "The Well Posed Problem" in *Papers on Probability Statistics and Statistical Physics* ed. Rosenkrantz, Reidel (Boston, 1983), pp. 131–149.

However, insofar as the logicist's view differs from that of the personalist's (for whom the equivalence relation reflects only a subjective judgment), logicists need to *motivate* their acceptance of a particular relation of congruence as a guide for choosing their probabilities. We have tried, therefore, to show that in certain circumstances the motivation stems from another principle, namely, the equiprobability of events that could evolve into each other. This principle states that, if probability is to be regarded as a physical parameter, it should be time-invariant.[19] Consequently, we have started again with a homeomorphism T and its iterates T^n, which describe the "evolution" of events, and showed that "$\cong$" could be defined relative to T. Similar ideas are used in other areas of physics.

As we have seen, however, one of the conditions that had to be satisfied by the above class of homeomorphisms was P_3, namely, for all $a,b \in S$ there will be a $T \in M$ and an n such that $T^n(a) = b$. This condition, as we have said, is rarely satisfied by physical systems, and in fact it can only happen in the case of systems with one degree of freedom. Therefore there is still a problem – how to motivate and justify the idea that two congruent sets must be equiprobable?

One possible way of avoiding this situation is by relaxing the requirement that for all a,b in S there be an n such that $T^n(a) = b$. Indeed, that is precisely how topological dynamics began. Let us review some of the possible relaxed requirements.

$\mathbf{A_1}$. For all $a,b \in S$ and for every $\in\, > 0$ there will be an n such that $T^n(a)$ will be in an $\in$-neighborhood of b. Systems that satisfy this condition are called *minimal* systems.

$\mathbf{A_2}$. A way of further relaxing this condition is to require that, for "almost" all $a \in S$, the *orbit* of a, that is, the set $\{a, T(a), T^2(a)\ldots\}$, will be dense in S.

$\mathbf{A_3}$. An even weaker condition is that if $A \in \mathbf{B}$ and $T(A) = A$, then A is either very small or very big. (This would imply that even small sets – but not very small – would cover the whole of S.) When the terms "very small" and "very big" are replaced by the measure-theoretic concepts of measure zero and measure one, respectively, we arrive at the concept of ergodicity.

Remember, though, that our project was to motivate the choice of a probability measure; therefore, the use of ergodicity in the "relaxed"

---

[19] The issue is the stationarity requirement that we discussed in Chapter 2.

construction is objectionable because it leads to circularity. The same is true of the second proposed relaxation, A_2, because it involves the term "almost," which presupposes a prior concept of measure.

Now, if this circularity does not bother us, we could continue with the Haar construction in ergodic systems. For example, $h(A,B)$ could be defined as the minimal n, such that $m(\cup_{i=1}^{n} T^i (A)\backslash B) = 0$. Provided that $m(A), m(B) > 0$ for the relevant m, the requirement that our system be ergodic will ensure that this definition is adequate. If $h(A,B)$ were not defined, that is, $\cup_{i=1}^{\infty} T^n (A)$ would not cover B, we would have $\cup_{i=1}^{\infty} T^n (A) = T(\cup_{i=1}^{\infty} T^n (A))$ but $m(\cup_{i=1}^{\infty} T^n (A)) \neq 1$. Continuing the Haar construction from this point is a relatively easy matter. However, the question of motivating m, as we remarked, cannot be answered in a noncircular way.

Although the first relaxation A_1 does not lead to circularity, it has a very limited range of physical applications. Another concept that does not lead to circularity is that of a transitive system. Such systems have points whose trajectories are dense. It turns out that if there are some points with this property, the set of such points is dense. Therefore, for all $a,b \in S$, there are $c,d \in S$ and $n > 0$ such that a is near c, b is near d, and $T^n(c) = d$. If we take this fact to be a partial satisfaction of P_3, we could continue with a relaxed Haar construction and define $h(A,B)$ to be the smallest n, such that $U_n T^n(A)$ is dense in B. Generally, there are interesting ways to use these topological concepts for better formulations of mechanical ideas. In the next chapter we shall come back to these issues. However, to use these ideas, we shall have to leave the strict Hilbertian eliminativist framework.

3.5. HOPF'S NOTION OF EQUIVALENCE BY DECOMPOSITION

The ideas concerning possible relaxations of the requirement P_3 took us far afield from Haar's original construction. P_3, as you may recall, was the requirement that a single group element will generate the whole group. Our ideas, on the other hand, concerned the conditions under which a single element will "almost generate" the group. The concepts of "almost generating the group" and an "approximate generator" were never developed as algebraic concepts. In this sense, we did indeed move away from the original construction.

In this section we shall continue our departure from Haar's original setting. We shall first define the concept of a nonwandering element and show that it is a generalization of the concept of an "approximate generator." Using this concept, we shall give a further relaxation of P_3. Finally, we shall state a theorem that gives necessary and sufficient conditions for there being a unique ergodic probability that has the right relation to the frequencies. These conditions are not very meaningful as means for classifying physical systems. However, the ideas behind the theorem may serve to introduce the reader to the vast literature on descriptive set theory. This literature has significant relevance to our concerns.

Let $\mathbb{S} = \langle S, B, T \rangle$ be a measurable dynamic system. A set $A \in B$ is a *wandering set* if whenever $i \neq j$, $T^i(A) \cap T^j(A) = \varnothing$. Let $W \subset B$ be a σ-ideal. (We shall think of W as the σ-ideal of negligible sets.) We shall say that $\mathbb{S}$ is *bounded* if all of the wandering sets are negligible. If $A \in B$ is nonwandering, for some i and j, $j \neq i$ $T^i(A) \cap T^j(A) \neq \varnothing$. Let $s \in T^i(A) \cap T^j(A)$. Suppose that $i > j$. That is, suppose that $i = j + r$. Because $s \in T^i(A)$, $s = T^i(s')$ for some $s' \in A$. Similarly, because $s \in T^j(A)$, $s = T^j(s'')$ for some $s'' \in A$. Hence, s' "comes back to its origin," that is, to $T^i(A) \cap T^j(A)$ after r iterations of the transformation T. Moreover, let $A' = T^{-i}(T^i(A) \cap T^j(A)) = T^{-i}(T^i(A)) \cap T^{-i}(T^j(A)) = A \cap T^r(A)$. Let $s \in A'$. (We may assume that A' is not empty because it is bounded.) Because $s \in A'$, by necessity, $s \in A$ and $s \in T^r(A)$. Hence, for some $s' \in A$, $s' = T^r(s)$. Hence, s itself "comes back to its origin" after r iterations. By the same argument, it is clear that $T^{nr}(s) \in A$ for every n. Hence, s is "infinitely recurrent." In fact, the above argument proves that W-almost every s is infinitely recurrent. That is, all but the members of a negligible set are infinitely recurrent. This argument is known as Poincaré's recurrence theorem.[20]

Definition. *Let* S *be a bounded system. We shall say that* S *is weakly ergodic iff, for all* $A \in B$, $T(A) = A$ *implies that* $A \in W$ *or* $S \setminus A \in W$.

Theorem. *Let* $s, s' \in S$. *Suppose that* S *is bounded and weakly ergodic (and separable). Let* $A \in B$ *be any "nonnegligible neighborhood" of* s. *Then, it is* W-*almost certain that, for some* n, $T^n(s') \in A$.

Because A can be as "small" as we want, we can paraphrase the theorem as saying that, when bounded weakly ergodic systems are concerned, almost every point comes arbitrarily close to every other point. Hence we can regard the theorem as stating that the requirement that S is a bounded weakly ergodic system is yet another relaxation of P_3.

---

[20] For a more comprehensive discussion of Poincaré's recurrence theorem, see Oxtoby's *Measure and Category*, Springer-Verlag (New York, 1971), pp. 65–70. Generally, Oxtoby's book contains many of the ideas discussed in this section.

However, what is special about this generalization is that we used neither probabilistic nor topological notions to define boundedness or weak ergodicity. In addition to references to the structure $\langle S, B, T \rangle$, the only other reference we made was to the σ-ideal W, which is taken to be a set of subsets of S having some closure properties. Observe, then, that if we had an equally abstract characterization of W, the above-mentioned theorem would be a genuine improvement in our attempt to characterize P_3. Indeed, an interesting idea in this direction was suggested by Hopf[21] in the following manner.

Definition. *Let* $A, A' \in B$. *We shall say that* A *and* A' *are equivalent by decomposition iff:*

1. $A = \cup_{i=1}^{\infty} A_i$, *where* $A_i \cap A_j = \varnothing$.
2. $A' = \cup_{i=1}^{\infty} A'_i$, *where* $A'_i \cap A'_j = \varnothing$.
3. *For every* i *there is an* n *such that* $T^n(A_i) = A'_i$.
Or, at least, that for some n
3*. $T^n(A_i) \backslash A'_i \in W$.

It easy to see that equivalence by decomposition is indeed an equivalence relation.

Definition. *A system* $S = \langle S, B, T, W \rangle$ *is Hopf-bounded iff, whenever* $A, A' \in B$ *are equivalent by decomposition,* $A \bigtriangleup A' \in W$.

Notice that the definition of Hopf-boundedness includes a reference to W; hence, it cannot be used as a solution to the problem of motivating the choice of negligible sets. At best, we can regard Hopf-boundedness as a condition that has to be fulfilled for the definition of W to be adequate. However, the concept of Hopf-boundedness does introduce a new generalization of Haar's construction. Haar's original idea was to define the equivalence of A and A' using congruence through T. The concept of equivalence through decomposition extends this idea. Even though T may not transform A to A', it may transform the different parts of A to the different parts of A'. In such a case, regarding A and A' as equivalent is an idea that has a considerable intuitive appeal.[22]

Using the concept of Hopf-boundedness we can now state the following result:

---

[21] See Hopf, "Theory of Measure and Invariant Integral," *Trans. Am. Math. Soc.* 34 (1932), pp. 373–393. Hopf actually used the term "equivalence by division."

[22] For further discussions, see S. Wagon, *The Banach–Tarski Paradox*, Cambridge Univ. Press (Cambridge, UK, 1986).

Theorem (Nadakarni).[23] *Let $\langle S, B, T, W \rangle$ be a weakly ergodic and Hopf-bounded system. There is a unique T-invariant ergodic probability measure m on B that is W-invariant.*

This theorem, as we said before, does not solve the problem of the physical interpretation of probability. However, the theorem has a genuine importance for both subjectivists and frequentists. For the former, the theorem we cited is the strongest representation theorem of which we are aware. For the frequentists, it constitutes a very abstract characterization of ergodicity, a property that guarantees, as the reader is well aware of by now, that an adequate definition of relative frequencies is possible.

The concept of equivalence through decomposition is one of the oldest issues that presented itself to the originators of measure theory.[24] In fact, it was through the study of this concept that non-Lebesgue measurable sets were first discovered. The originators of measure theory believed that every set of the reals was measurable. They also hoped that the Lebesgue measure would preserve equivalence through decomposition. However, Tarski, who built on earlier work by Lebesgue, Vitali, Banach, and Hausdorf, showed by using the axiom of choice that σ-additive measures that are defined for every subset of the reals do not preserve equivalence through decomposition. The sphere, for example, can be shown to be equivalent by decomposition to a proper subpart of itself. That result was the beginning of what is now known as *descriptive set theory*. We shall return in Chapter 5 to the ideas presented in this section.

3.6. MAXWELL'S CONCEPTION OF THE ERGODICITY ASSUMPTION

Another point worth mentioning is the relation between the Haar construction and an idea suggested by Maxwell in the following passage:[25]

---

[23] See Nadakarni, "A Descriptive Characterization of Ergodic Systems," *Sankhya* 45 (1983), pp. 271–287. See also Nadakarni, "Descriptive Ergodic Theory," *Am. Math. Soc.* 94 (1989), pp. 181–206.

[24] For a discussion on the history of this topic, see the introduction to Wagon's book, ibid. See also Moschovakis, *Descriptive Set Theory*, North Holland (Amsterdam, 1981), pp. 1–9.

[25] See Maxwell, "On Boltzmann's Theorem on the Average Distribution of Energy in a System of Material Points," *Trans. Cambridge Phil. Soc.* 12 (1879), p. 547. For an interesting discussion on this passage, see *Creating Modern Probability*, ibid, p. 102.

If we suppose that the material particles, or some of them, occasionally encounter a fixed obstacle such as the side of the vessel containing the particles, then, except for special forms of the surface of this obstacle, each encounter will introduce a disturbance into the motion of the system, so that it will pass from one undisturbed path into another. The two paths must both satisfy the equation of energy, and they must intersect each other in the phase for which the conditions of encounter with the fixed obstacle are satisfied, but they are not subject to the equation of momentum. It is difficult in a case of such extreme complexity to arrive at a thoroughly satisfactory conclusion, but we may with a considerable confidence assert that except for particular forms of the surface of the fixed obstacle, the system will sooner or later, after a sufficient number of encounters, pass through every phase consistent with the equation of energy.

If this idea could be rigorously defended, it would yield another way of introducing physical content into the requirement P_3. Ordinarily, we assume that the collisions with the walls of the container are perfectly elastic. That is why we can represent the walls of the container as fixed parameters, which we refer to as the boundary conditions. Given the Hamiltonian and these conditions, the trajectories of the system are fixed. However, we can interpret Maxwell as saying that we should regard the boundary conditions as being only *nearly* fixed, a rather natural idea because the boundaries of real vessels do not have a precise time-invariant shape. From a mathematical point of view, this idea has far-reaching consequences. Rather than representing a mechanical system as a fixed collection of trajectories, we must represent it as a class of such collections. As time goes by, the slightly changing boundaries will switch the system from a trajectory that is a member of one collection in this class to a trajectory that is a member of another collection. This idea gives rise to a new interpretation of the conditions under which P_3 will be satisfied: Given that the current state of the system is s, and given sufficiently many disturbances caused by collisions with the wall of the vessel, the system will eventually switch to the trajectory of a given point s'. And, if this "patched up" trajectory could fill the whole space, we could define the probabilities as the time averages that are relative to this trajectory.

The virtue of this suggestion is that it yields a less idealized picture of mechanical systems. Indeed, there is much to be said against the idea that the walls of the vessel should be taken as completely fixed parameters. However, the dynamics of the walls is itself governed by a completely deterministic process. Therefore, rather than viewing the effect of the walls as a random source, we should aspire to correct our formulation by introducing a more inclusive Hamiltonian that will express the evolution of the particles as well as the evolution of the walls. Moreover, by regarding the effect of the walls as a random source, our account of the probabilities will become circular because randomness is a probabilistic concept. This way of reformulating condition P_3 will not provide us, then, with a solution to the foundational issues that have preoccupied us throughout this chapter.[26]

Notice that there are other ways to read Maxwell's statement. It is possible to interpret him as saying that the effect of the collisions with the walls should not be regarded as a random source but a source of instability; what we mean is that any deviation from a precise description of the walls of the container will yield completely different results. In the next chapter, we shall explore this and related arguments. Incidentally, we hope that the discussion in this section will deter the reader from confusing the view that Maxwell expressed in the above passage with a statement of the ergodic hypothesis in its simple erroneous form. We shall return to the discussion of Maxwell's views in Chapter 4.

3.7. THE ELIMINATION OF PROBABILITY

Let us end this chapter by canvassing how to obtain a Hilbert-style or a Quine-style elimination of probabilities from the Haar construction.

Let S be a sentence of the form, "The probability of a system M assuming a state $s \in A$ is p." The equivalent but probability-free reduction sentence would be, "The relative frequency in which M assumes states that are in A is p." As long as there is no ambiguity concerning what we refer to as the relative frequency, the construction of a Haar measure will ensure that the reduction sentence is unambiguously defined. This will be true for both the discrete and the continuous cases. As we have seen, when M is a system with one degree of freedom, even the relative frequency measure can be operationalized because we can actually measure the time averages.

Another way of thinking about the result of this chapter is as a demonstration that, in the case of systems with one degree of freedom, the "geometry" of $\langle S, B \rangle$ as codified by the relation of congruence "$\cong$" is sufficient to construct a probability measure on $\langle S, B \rangle$, and uniquely

---

[26] Notice that the question of the possibility of a "space-filling" line will remain a problem even if the above objections are met.

149

so. Speaking somewhat loosely, this conclusion may be restated as the following: The probabilistic truths of $\langle S,B \rangle$ can be deduced from the geometric ones. More precisely, we may conjecture that the "probability theory" of $\langle S,B \rangle$ may be a conservative extension of $\langle S,B,\cong \rangle$.[27] This reduction, however, cannot be extended for more degrees of freedom than one. It is worthwhile to observe that, as the concept of the limit of relative frequencies involves a limit operation, it may not always be harmonious with the Hilbertian finitist ideology. Only when the convergence to the limit is uniform can this reduction be construed in a manner that is acceptable to the finitist.

[27] A theory T is a conservative extension of the theory T' iff T includes T' and any semantic consequence of T is a semantic consequence of T'.

4

Measure and Topology in Statistical Mechanics

4.1. THE NATURALIST CONCEPTION OF PHYSICS

In the previous chapters we were occupied with the dual theme of the centrality and the problematic character of the frequentist's conception of probabilities in the context of SM. On the positive side, the frequentist's conception links the probabilities directly with the laws of physics, because the calculation of frequencies is effected through an integration along the trajectories formed by the successive solutions of the equations of motion. However, because there are in general many such trajectories, to each there corresponds a different value of the frequencies.

We saw in Chapter 3 how, in the case of a system with a single degree of freedom, there is only a single possible trajectory that the system may follow, and hence a single way of calculating the frequencies. In this case, which is the "best-case scenario" from the frequentist's point of view, it is possible to *reduce* the probabilities to relative frequencies. However, we have seen that this case is exceptional from a physical point of view, and so we have determined it necessary to reformulate the frequentist's standpoint.

One reaction to the failure of the simple frequentist's conception is the idea that, rather than calculating the exact frequencies along individual trajectories, we should calculate the average of the frequencies among the members of a set of trajectories. Having done that, we may ask whether these averages are different relative to the choice of different sets of trajectories (ergodic systems have the property that the different sets will yield the same averages). This idea was developed mathematically by von Neumann and is related to the subjectivist's interpretation of ergodic theory that we introduced in

Chapter 2.[1] According to this interpretation, the averages are the expectation values of the frequencies relative to a subjectivist probability measure.

In this chapter we shall present another reformulation of frequentism, one that moves in the opposite direction from subjectivism. Rather than looking for the conditions under which the average frequencies in different sets of trajectories will be the same, we will look for conditions that guarantee that every trajectory will pass near each of the points in the phase space. Using this condition, which is known as *transitivity* or *regional transitivity*, we may hope to ensure that the time averages that are relative to different trajectories are approximately the same. In other words, we may attempt to solve the problems of the frequentist's conception of probability by using the notion of approximation. As we shall see, this generalization of the frequentist's conception can be seen as an eliminativist's conception of SM probabilities motivated by a form of naturalism with respect to physics.

Let us begin the exposition of this eliminative attempt with a formative argument that defines the spirit of the "naturalist" program. The argument begins with an affirmation of an idea, one of the basic commitments of classical mechanics, that the equations of motion, together with the initial conditions of a mechanical system, determine the trajectory of the system uniquely. This means that, given the equations and the exact initial conditions, there are no further degrees of freedom. Both the past and the future states of the system can be determined (in principle) from these two factors alone. In this sense, the argument continues, the trajectory structure of the system (that is, an assignment of a unique trajectory to any possible initial state) contains the totality of physical facts about the system. After all, what other facts could there be beyond those that determine all of the physical properties of the system for any point in time? Therefore, the argument concludes, what cannot be derived from the trajectory structure alone cannot be a physical fact; at best it may be a "formal" fact, that is, a fact that is an artifact of the system of representation.[2] The view that physics should be reformulated to avoid formal

---

[1] See von Neumann, "Proof of the Quasi Ergodic Hypothesis," *Proc. Nat. Acad. Sci.* 18 (1931). The operation of averaging over the frequencies of a whole set of trajectories can be interpreted as a result of ignorance with respect to the identity of the right trajectory.

[2] The term "formal fact" will not be precisely defined here, but we trust that the sentiment behind it is sufficiently clear. When a physical theory is interpreted in a mathematical model, the adequacy of the model depends on the interpretation of the

facts as much as possible might be labeled naturalism with respect to physical facts, for lack of a better term.[3]

Now, probabilities are usually introduced in classical physics as assignments of "weights" to sets of possible trajectories. When the number of trajectories is finite, these weights are simply the relevant ratios. But when physical systems are described using "continuous" parameters, it can be proven that no single probabilistic assignment can be *derived* from the trajectory structure itself. And if the totality of physical facts is not sufficient to determine the probabilities, says the naturalist, the probabilities themselves cannot constitute physical facts. Therefore, physics should be reformulated or reconstructed so as to avoid probabilistic concepts.

As seductive as this extreme form of naturalism may seem, there are many reasons to criticize it, some of which are independent of its consequences for the theory of probability. One important objection is that the trajectory structure itself does not determine the limit behavior of the system. That is, the trajectory structure itself does not determine, for example, whether or not a trajectory "converges" in the limit to a point. This objection is quite serious because in many cases the adequacy of the mathematical models that physicists use depends on assumptions involving limit behavior.[4] Therefore, we believe that a workable naturalist approach must reconcile itself with the idea that the limit behavior of a system is a physical fact. This version of naturalism, or, rather, the elimination program of probabilities that it inspires, seems to us to be a serious foundational standpoint that merits special attention.

When the concept of limit behavior is admitted to the realm of

theory as a whole. One of the things that can happen as a result is that, whereas the modeling is generally adequate, it is not parsimonious; that is, a "slimmer" model would have been adequate as well. A naturalist takes this to mean that the less economic model contains "formal" facts that should be eliminated.

[3] Naturalism, then, like instrumentalism, inspires the desire to put ordinary physics through an "ontological diet." But, while the instrumentalist is interested in the diet per se, the naturalist is interested in it only as means of preserving the "physical core" of our theories. Therefore, the naturalist may tolerate added assumptions involving objective dispositions if he or she is convinced that they have a physical basis. This tolerance for objective dispositions places the naturalist in opposition to the subjectivist, for whom the elimination of objective dispositions is of primary philosophical importance.

[4] Take as a simple example the way in which we model the time axis. If we did not tacitly include a point of infinity (which in this context functions as a limit point), there would be many nonstandard ways to model time.

physics, the result is a mathematical model of the trajectory structure endowed with some topological structure. Within such a structure, the concept of the distance between states can be defined. By using this concept we can then define the notions of approximation and stability. In turn, these two concepts can be used to define a host of other important concepts in an extremely economical and elegant way. In particular, one may use these concepts to attempt a probability-free reconstruction of SM. More concretely, the reconstruction that is discussed in this chapter centers around the idea that we may justify the introduction of statistical methods into SM by the assumption that the time averages can be calculated relative to a trajectory that *approximates* every point in the phase space, rather than requiring that *most* trajectories will yield the same time averages, where "most" is understood probabilistically. The replacement of probabilistic concepts with approximative ones changes the content of SM. The approximative second law explains the passage to equilibrium as a consequence of the dynamic properties of the system (i.e., because every trajectory is approximately "maximally entropic"). This property can be formulated as a disposition of each individual trajectory. According to the probabilistic version, on the other hand (especially when this version is given a subjectivist formulation), the second law is formulated to be a property of an ensemble of systems. This approach explains the passage to equilibrium in terms of the average behavior among the members of the ensemble. Consequently, the demand for an approximative reconstruction of SM can be motivated by the general philosophical view that the knowledge of average behavior is not sufficient for explaining the behavior of individual cases.

At this point some readers may wonder why we call this paraphrasing a reconstruction, because it is rather obvious that the original probabilistic version and the reconstructed "approximative" version are neither synonymous nor equivalent in any simple sense. That is, one may rightly ask: What is the feature of the relation between the original theory and its reformulation that allows us to talk about a reconstruction in the first place? The answer to this question is based on a deep mathematical fact, namely, the Erdös–Sierpinski duality principle.[5] This

---

[5] See Oxtoby's excellent *Measure and Category*, Springer-Verlag (New York, 1970), pp. 74–78. The book also contains a very elegant introduction to the concepts that we shall introduce in this section.

principle is a theorem that proves that probabilistic and topological approximative concepts have a similar underlying structure that guarantees that they are equivalent in a certain sense. The equivalence, in turn, justifies the term reconstruction.

Our analysis follows roughly the historical development of the science of mechanics. The mechanics of Newton was developed as a geometric theory capable of describing, in principle, the evolution of systems of any complexity. Towards the end of the nineteenth century, however, it was felt that neither the behavior of very orderly systems, such as celestial systems, nor the evolution of seemingly random systems, such as SM systems, were adequately accounted for by the classical Newtonian assumptions. Therefore, in this period two new theories emerged: topological dynamics, which attempted to analyze systems with few degrees of freedom, and ergodic theory, whose goal was to analyze SM. Topological dynamics is concerned with concepts such as limit behavior, stability, and approximability. According to this theory, all of these concepts are analyzed as dispositions of individual systems. Ergodic theory, on the other hand, was developed as a means to analyze aggregates or ensembles of systems. The concepts of equilibrium and entropy were meant to apply to such ensembles. In one sense, then, topological dynamics may be characterized as an analysis of the dynamic dispositions of individual systems and ergodic theory as the theory of aggregate behavior. Viewed in this light, topological dynamics and ergodic theory are not just different mathematical structures; they represent two different foundational attitudes. In particular, it is possible to regard topological dynamics as the framework in which a nonprobabilistic reconstruction of SM can take place. It is this point of view that most concerns us here.

To the best of our knowledge, the topological dynamic point of view has never been presented systematically as a foundational standpoint in the philosophical literature on the subject of mechanics. This situation is surprising, because some related ideas were already discussed at the end of the nineteenth century by Poincaré, Lyapunov, Hadamard, and Duhem. In the 1920s and 1930s, these ideas were further developed mathematically by Birkhoff, Morse, Hopf, and Hedlund. More recent contributors to this mathematical tradition are Abraham, Arnold, Sinai, Anosov, and Smale.

155

Probabilistic concepts are defined in measure spaces, whereas approximative concepts require topological spaces.[6] Both measure spaces and topological spaces are obtained by an abstraction from geometric spaces such as the Euclidean spaces E, which have both a natural measure and a natural distance metric. But Euclidean spaces have other properties as well. For example, they possess a notion of direction or orientation that is a property of neither abstract measure spaces nor of abstract topological spaces. Only spaces that have measure, metric, and orientation are proper geometric spaces.[7] Let us begin, then, with an account of some of the elementary properties of metric spaces. We have already had ample opportunity to become acquainted with the properties of the measure space $\langle S, B, m \rangle$.

A metric space[8] is an ordered pair $\langle S, d \rangle$, where S is a set and the metric $d: S \times S \to R$ is a function that obeys:

a. $d(s, s) = 0$ for all $s \in S$.
b. $d(s, s') = d(s', s)$.
c. $d(s, s'') \leq d(s, s') + d(s', s'')$.

A point s is called a *limit point* of the set $A \subset S$ iff there is a sequence of points in A that converges to s, that is, points whose distances from S converge to zero. The closure $\tilde{A}$ of a set A is obtained by adding to A all of its limit points. A set A is called *dense* if $\tilde{A} = S$. A set A is called *nowhere dense* if the complement of its closure contains a dense set. Subsets of S are divided into sets of the first and second categories. A is called a *set of the first category* if it can be presented as a countable union of nowhere dense sets. If a set is not of the first category, it is of the second category. Sets of the first category are regarded as "small sets." A property that holds for a set of the second category is called a *generic property*.

Loosely speaking, the metric fixes the concept of a distance between

---

[6] For an introductory text on these concepts, see Oxtoby, ibid.

[7] Only in geometric spaces can we attribute to sets a "shape" that we may use to classify them according to their level of complexity. Therefore, the discussion on the notion of "coarse graining" presupposes that we have already defined all of the necessary geometric notions.

[8] A more general topological space is a pair $\langle S, A \rangle$, where S is a set and A is a collection of "open sets." If A has a countable basis, $\langle S, A \rangle$ is called metrizable because it is possible in that case to prove that a metric d can be defined on $\langle S, A \rangle$.

points,[9] and the concept of measure defines the volume of a set of points. This way of speaking, however, can be misleading. A measure can be interpreted as volume only in spaces that already have a metric structure. In abstract spaces, the concept of volume does not make much sense. Therefore, in the abstract, measures should be interpreted as attributing "weights" to sets of points, that is, they should be interpreted probabilistically.[10] Mathematicians usually do not adopt this point of view, because they regard completely abstract measure spaces to be superficial from a mathematical point of view. They think that topology defines structures whose properties remain invariant with respect to "stretching," and that the measure comes "on top" and fixes the concept of volume. There is a complementary view held by some statisticians who posit that probabilistic spaces "come first," and the metric structure may be obtained by the addition of concepts that define the notion of approximation in the space. In our opinion, neither of these views seems very convincing, and our advice then is to regard both the concept of measure (when interpreted probabilistically) and the concept of an abstract topological space as independent autonomous concepts. The concept of volume, then, should be defined only in those contexts where both structures are present. In such contexts, the volume measure is obtained through constraints imposed on the measure by the metric structure. One common constraint is regularity: A probability measure is regular if it assigns positive values to open sets. Open sets are intended to have a positive measure because they possess a nonempty interior. (The interior of a set is the set of points in the set that are surrounded by a neighborhood that is also in the set.) In compact metric spaces, the Lebesgue measure is simply a regular measure; in such a case, the concept of regularity leads to uniqueness.

There is a noteworthy analogy between topological concepts and

---

[9] In the space of functions $f: X \to X$, the distance between the functions f and g is often defined as:

$$d(f,g) = \sup_{x \in X} d(f(x), g(x)) = \sup_{x \in X}$$

Relative to this metric we can define open ε-neighborhoods of functions, and relative to such neighborhoods we can talk about the concept of genericity in the function space.

[10] The concept of the volume of a set of points presupposes the topological concept of a connected set that cannot be defined in an abstract measure space. Without this notion we can only talk about measures as attributing "weights." The concept of weight is essentially a probabilistic concept.

157

measurable concepts. The analogy is due in part to the fact that both the set of subsets of the first category and the set of subsets of measure zero (or the null sets) form a σ-ideal. Recall that W is a σ-ideal if the following conditions hold:

a. If, for all $\{A_n\}_{n=1}^{\infty}$, $A_n \in W$, then $\bigcup_{n=1}^{\infty} A_n \in W$.
b. For all $A \in W$ and all $C \in B$, $C \cap A \in W$.

These properties are clearly satisfied by both first category sets and by the null sets. Another point of similarity is that both the set of null sets and the set of first category sets include the countable and finite sets. But, more generally, Sierpinski[11] and Erdös[12] proved the following metatheorem:[13]

Theorem. (Erdös–Sierpinski 1934–1943). *Let* P *be any proposition involving solely the notions of measure zero, first category, and pure set theory. Let* P* *be the proposition obtained from* P *by replacing "null sets" with "sets of first category," and vice versa, wherever they appear. Assuming the continuum hypothesis,* P *and* P* *imply each other.*

This theorem, which is called a duality theorem, defines the extent to which probabilistic and metric concepts are interchangeable. This phenomenon is quite extraordinary, because in most structures that have both a natural measure and a natural metric the respective notions of small sets do not coincide! The dramatic extent to which they do not coincide can be gathered from the following theorem: *The real line can be partitioned into two sets A,B such that A is a null set and B is a set of the first category.*[14]

We remarked earlier that physical theories are usually modeled in mathematical structures that have both a topological structure and a measurable structure. By using these structures it is possible to define both the notion of convergence and the notion of an integration on the space. These notions are necessary for a precise formulation of our physical theories. In the next section we shall begin with an example of the way in which a physical system is given a precise mathematical modeling.

---

[11] Sierpinski, "Sur la dualite entre la premiere categorie et la mesure nulle," *Fund. Math.* 22 (1934), pp. 276–280.
[12] Erdös, "Some Remarks on Set Theory," *Ann. Math.* 2(44) (1943), pp. 643–646.
[13] See Oxtoby, ibid.
[14] See ibid, p. 5.

4.3. TOPOLOGICAL DYNAMICS AS A FOUNDATION
FOR MECHANICS

The laws governing the evolution of physical systems are formulated as a system of differential equations that describe the functional relations between the different parameters.[15] These relations determine the evolution of the system. But the equations give us even more information about the system: At each moment, the partial derivatives determine the rate and the direction of change relative to each of the parameters. Therefore, it is natural to represent the system of equations as an "action" on a vector field. The functional relations between the parameters determine the ways in which the states of the system change in time, and the partial derivatives represent the rate of change as a vector attached to each point.

Let us take an example that we can refer to continuously. Let M be a damped oscillator. An example of such a system is a pendulum swinging toward a rest point p as a result of a loss of energy due to friction. A description of the change of the angle of M as a function of time can be given by the equation

$$d^2q/dt^2 + g\,dq/dt + w = 0, \qquad (*)$$

where g is a "damping coefficient" and w is a constant that is determined by the angular frequency. By analyzing (*) it becomes clear that the action it describes is a spiral in the position and momentum space. (We can add a temporal z-axis and obtain a three-dimensional spiral.)

The assumption that systems such as M are deterministic is expressed by the uniqueness of the solutions of the system of the differential equations (*). In turn, the latter property implies the fact that there is only one "line of force" going through each point in the vector field. Because of this form of determinism, for 200 years, the main efforts of mathematicians and physicists were invested in devising methods for quicker and more efficient and exact solutions of differential equations. From the seventeenth to the nineteenth centuries, remarkable progress was made in this direction; many methods were developed to find the "solutions" to the system, given its initial conditions. This period is considered the "classical period" in the development of the theory of differential equations.[16]

---

[15] For an introduction to the subject, see, e.g., Abraham, *Foundations of Mechanics*, Benjamin Cummings (Mass, 1978).

[16] For a brief historical account, see ibid, pp. 1–4.

Only toward the end of the nineteenth century did the inadequacy of the classical approach become obvious. Because we usually do not know the exact initial conditions of the system that we are investigating, the quantitative methods are informative only if we can safely assume that any two trajectories passing through an "interval of uncertainty" will remain "within bounds" from each other. This assumption is not always satisfied. In particular, when systems with many particles (or "degrees of freedom") are concerned, we can be virtually certain that no such nontrivial bounds will be found. Therefore, the end of the nineteenth century saw the beginning of two new approaches: Poincaré began to develop topological dynamics to investigate the conditions under which the behavior of mechanical systems may be approximated by imprecise measurements,[17] and Maxwell and Boltzmann initiated the ergodic theory that investigates with probabilistic tools the behavior of systems with many degrees of freedom.[18]

Therefore, topological dynamics and ergodic theory were developed as attempts to expand traditional mechanics in opposite directions: The former was designed to describe the approximate behavior of systems with few degrees of freedom and the latter to investigate the "stochastic" behavior of many-particle systems. Beyond the immediate original intentions, though, the two theories can be seen as alternative foundations to the theory of the behavior of mechanical systems. In its purest form, ergodic theory is an attempt to describe the average dynamics using probabilistic concepts, whereas topological

[17] Poincaré first published his results in *Les Methods Nouvelles de la Mecanique Celeste* (published in 3 volumes between 1892 and 1899 by Gauthier Villars). His investigations were carried on afterward by Lyapunov and Birkhoff. As we remarked earlier, another reason for the development of topological dynamics was the need for a clear formulation of the concepts of limit behavior. Regarding the pendulum, for example, we need to express the fact that, although the spiral describing the course of the evolution of the pendulum will never reach the rest point p (that would violate the uniqueness of the trajectory assigned to point p), it will nevertheless converge to p. This relation also cannot be clearly formulated within the classical framework. According to the classical approach, all of the information concerning a physical system is determined by the trajectory that the system follows. But because the limit point of the trajectory (if it exists) is not on the trajectory itself, the relation between the trajectory and its limit point is located outside the concerns of the classical approach. Even the formulation of this relation necessitates an extension of the classical framework.

[18] Boltzmann did, of course, treat systems with only a few degrees of freedom in his writing. However, his concept of entropy, for example, makes sense only for systems with many particles.

dynamics attempts to reconstruct mechanics in "qualitative" approximative terms. There are, to be sure, many "intermediate" versions as well.

Due to their structural similarities, both topological dynamics and ergodic theory can be formulated within a single abstract dynamic framework. This abstract theory requires a space $\langle S,B,W,T \rangle$, where S,B, and W are, as before, a set, a σ-algebra, a σ-ideal, and a 1–1 "measurable" transformation. The abstract structure $\langle S,B,W,T \rangle$, then, is the skeleton of both ergodic theory and topological dynamics, which means that, in essence, both theories can be developed within the abstract framework when the σ-ideals are comprised of the null sets and the sets of first category, respectively.

Let us briefly describe the transition from the abstract theory to ergodic theory and to topological dynamics. We shall begin with a few useful notions that can be defined abstractly.

1. The limit set of A, which we shall denote $\lim(A)$, is defined as $\bigcap_{n=1}^{\infty} \overline{\bigcup_{i=1}^{n} T^i(A)}$. (A line over a set signifies the closure of the set.)

2. Recall that the average of stay in A (or the time average of A, as it is usually called) is the quantity $\lim_{n \to \infty} 1/n \, \Sigma_{i=1} X_A(T^i(s))$, where X_A is the characteristic function of A. Notice that this is a purely set-theoretic definition.

3. A set $A \in B$ is a wandering set if $T^i(A) \cap T^j(A) = \emptyset$ for all i and j. This property, too, is a set-theoretic property.

4. A system is closed iff all of the wandering sets are "small" (i.e., members of W).

5. A system is nondissipative iff $T(A) \in W$ implies that $A \in W$.

6. Given $\langle S,B,W,T \rangle$, we can decompose S into maximally invariant subsets,[19] that is, a collection of measurable subsets A_i such that:
   a. $T(A_i) \backslash A_i \in W$.
   b. A_i is maximal in that respect.
   c. $A_i \cap A_j = \emptyset$.
   d. $\cup_i A_i = S$.

Systems for which the decomposition yields a single set A such that $S \backslash A \in W$ are called *indecomposable*. Ergodic systems, or systems for

---

[19] For the origin of this construction, see Oxtoby, "Ergodic Sets," *Bull. Am. Math. Soc.* 58, pp. 116–136.

which $A = T(A)$ implies that A is either a null set or a set of full measure, are indecomposable.[20] The most important feature of ergodic systems is that the time averages are the same, relative to all but a null set of initial states. Regarding nonergodic systems, the possibility of decomposing them into maximally invariant subsets provides a justification for treating each of the components as if they were separate ergodic systems. This construction justifies the idea that nonergodic *measures* should be regarded as weighted averages of ergodic measures.

Although the ergodic theory was developed as the abstract theory of conservative systems with many degrees of freedom, it can apply to systems with just a few degrees of freedom as well. Such systems could be ergodic if, for example, their boundary conditions are sufficiently irregular. Ergodic theory is sometimes extended to nonconservative but nondissipative systems as well.[21] But, although the idea of an ergodic theory of dissipative systems is possible, it is hardly ever developed.[22]

The framework of ergodic theory, then, defines a foundational attitude to general mechanics. According to this approach, what must be added to the classical trajectory structure is the concept of the measure of a set of trajectories. By using this concept, we can investigate the average behavior of the trajectories. Because of the preoccupation with averages, such an investigation is more appropriate to systems with a large number of degrees of freedom.

The ergodic theoretical approach may be given different interpretations that can in turn be divided under objectivist and subjectivist headings. For the subjectivist, the ergodic theorem proves that the choice of an ergodic measure as a measure of our degrees of belief will

---

[20] The concept of ergodicity can also be defined for transformations that do not preserve measure.

[21] We expanded on this topic in Chapter 2.

[22] To be sure, it is possible to extend ergodic theory in this direction as well. One way of doing so is to express the dissipation as the time-dependent loss of areas of accessibility. More precisely, dissipative systems can be characterized as having transition probabilities that are expressible by a time-dependent conditional probability function that concentrates in the limit on a set that was initially a null set. This construction, however, blurs the distinction between an impossible event, a set that is initially null and a set that becomes null because of the dissipation. These distinctions are a part of the content of SM. Dissipation is the "shrinkage" of the number of possibilities and not only an increase in the number of the null sets (see Chapter 2).

lead us to expect that the average limiting behavior will be the same *m*-almost always. Therefore, if the average time spent in the set of the states compatible with macroscopic equilibrium is very large relative to certain initial conditions, we can be practically sure that it will be large *m*-almost always. But let us not forget, warns the subjectivist, that the choice of an ergodic probability cannot be ultimately motivated in and of itself. The objectivist may also reason along more or less the same lines, but when the subjectivist abandons the prospect of justification, the objectivist will persevere, insisting that there is an objective reason for calling a system "stochastic." This heroic stance, however, in many cases ends in circularity, when stochastic processes are defined as ergodic processes. Other, more careful objectivists, on the other hand, rely at this point on topological characterizations of stochasticity. We will be in a better position to evaluate this strategy once we have reviewed the concepts of topological dynamics. Let us, then, see how topological dynamics was developed as an abstraction from the theory of differential equations.[23] In doing so, we shall make extensive use of the example of the damped oscillator.

As the reader may recall, when we analyzed the equations of motion that apply to the damped oscillator, we realized that the action of the equations could be described as a spiral converging to a point. Poincaré discovered this fact and went on to ask another question: How many different types of singular points are there in spaces of two dimensions? He found that there are only five qualitatively different types of singular points! The fact that all of the infinitely many sets of equations that give rise to singular points can be so neatly classified was one discovery that led to the basic "qualitative philosophy" of topological dynamics, namely, it brought about the claim that the most important things we can know about physical systems are the types of singularities that they possess. (Unfortunately, classifying curves in space with dimension higher than 2 is much more complicated.) This point of view justifies the practice of abstracting from the vast amount of information that the equations of motions contain, by merely noting the type of singularities to which the action of the equations lead.

---

[23] For an introduction to topological dynamics, see Szego and Bahtia, *Stability Theory for Dynamical Systems*, Springer-Verlag (New York, 1970). For an elementary presentation of the modeling of physical systems by differential equations, as well as other topics related to topological dynamics, see Berge et al., *Order Within Chaos*, Wiley (New York, 1984).

Therefore, in the case of the damped oscillator, for example, we can merely state that its phase space has only one stable, "focal" singular point. That is, even if we know the approximate location of the system, we can be sure that the error in our measurements will not be magnified as the system evolves in time, making our predictions increasingly inadequate. This feature is also true for other cases of stable convergence, but it is not true for other types of singular points. In some cases, a slight perturbation will lead to a very rapid divergence. Note that there are four different concepts of stability of motion.[24] Their definitions are not equivalent.

1. A trajectory o is stable in the sense of Poisson if nearby trajectories will return arbitrarily close to o infinitely often.
2. A trajectory o is stable in the sense of Lagrange if nearby trajectories will never diverge too far from o.
3. A trajectory is stable in the sense of Lyapunov if the limit behavior of o is the same as the limit behavior of other trajectories that intersect a neighborhood of the limit set of o.
4. A system is structurally stable if a "simultaneous perturbation" of all of its trajectories will yield a system of a similar type.

Instability is the absence of stability in any of the above senses.

The point p is called a *point attractor* because it attracts a large set of trajectories. The set of points on these trajectories is called the *basin of attraction*. There are other types of attractors that may be classified by their dimensions. The point is the lowest in this hierarchy, because its dimension is zero. Next we can find the limit cycle. There are two ways in which a system may have a limit cycle: One way is stable and the other is not. For every n there are attractors of dimension n that are called "limit tori."(A torus of dimension n is the product of n cycles.) There are other types of attracting sets that are not homeomorphic to tori; these are called *strange attractors*. Such attractors may have a noninteger dimension called the *fractal dimension*.

One of the most remarkable theorems in topological dynamics that, in a sense, justifies the whole enterprise of a qualitative classification of dynamic systems is Whitney's theorem.[25] The theorem states that any

---

[24] For formal definitions and a discussion, see, for example, Szego and Bahtia, ibid. pp. 56–114.
[25] See Whitney, "Regular Family of Curves," *Ann. Math.* 34 (1933), pp. 244–270.

system that does not possess singular points is homeomorphic to a system of parallel lines. This remarkable result explains why we can extract from the wealth of information contained in the differential equations only the "bare minimum" concerning the limit behavior and still maintain a meaningful description of the system.

All of the complex concepts and results of topological dynamics can be formulated in a system $\langle S,d,T \rangle$, that is, in $\langle S,\mathsf{B},W,T \rangle$, when W is interpreted as the σ-ideal of sets of the first category and S is required to be metrizable. For example, the set of points $p_0, p_1, \ldots$ converges to p iff $d(p_n, p) \to 0$ as $n \to \infty$. The convergence is spiral if, for all p_n, the distances $d(p_n, p)$ repeatedly increase to a local maximum and then decrease to a local minimum. The behavior of T is Lagrange-stable in p iff, given that $d(p, p')$, there is a δ such that $d(Tp, Tp')$.

In topological dynamics, indecomposable transformations are called *minimal*. The definition of a minimal system bears a strong resemblance to the definition of ergodic systems – a system is minimal iff, for all $A \in \mathsf{B}$, $T(A) = A$ implies that $A = S$ or $A = \emptyset$. A weaker analog of ergodicity is transitivity. T is transitive if $T(A) = A$ implies that $A \in W$ or $S \backslash A \in W$. Because of the analogy between minimality and ergodicity, one would expect that if T were minimal, *every* trajectory would have the same "time averages" (i.e., the average stay would be the same relative to every initial condition). This, however, is known to be false.[26] This negative result should serve as a reminder that the duality principle does not guarantee a complete resemblance between ergodic theory and topological dynamics. We should always be on guard and should never take the analogy for granted. The duality principle is based on a nontrivial mathematical fact and not on a literary simile that can be extended at will.[27]

Let us sum up this outline of topological dynamics and ergodic theory by saying that the evolution of a mechanical system in its phase space is usually described as an action of a set of differential equations on a vector field. Such a description includes a vast amount of information about the system. It is the purpose of the different foundational approaches to mechanics to construct more abstract mathematical

---

[26] R. Mane, *See Ergodic Theory and Differentiable Dynamics*, Springer-Verlag, (New York, 1983), p. 134.
[27] Another important difference between ergodocity and minimality is that, generally, nonminimal systems cannot be decomposed into minimal components (a minimal component is a closed invariant set).

descriptions that include only an "essential selection" of the facts. The classical approach describes mechanical systems in a deterministic trajectory structure, that is, a structure that assigns a trajectory to any initial state. The trajectory structure abstracts from the differential equations the property that at any time they have unique solutions. However, this structure is not rich enough to formulate the assumptions needed for an efficient description of both celestial systems and SM systems. There are two ways of extending the classical trajectory structure. In both cases, the extensions are obtained through further abstractions from the theory of differential equations. The first extension requires the addition of a notion of distance between states. With this notion we can define the concepts of limit behavior, approximation, and stability. This is known as the *topological dynamic approach* and was developed originally to solve problems in celestial mechanics. The second possibility is to define the concept of a measure of a set of trajectories. Using the measure, the concepts of time averages and space averages can be defined. This is the ergodic approach, which is by now the standard approach to SM.

It is worth thinking about topological dynamics and ergodic theory as rival foundational approaches to the science of mechanics as a whole and not only as particular formulations of celestial and statistical mechanics, respectively. Topological mechanics should be presented as an approach that insists on the explanatory significance of the evolution of individual trajectories. Ergodic theory, on the other hand, analyzes the average behavior among an aggregate of different trajectories.[28] (Of course, physicists do not usually adopt just one of these approaches and remain allied solely to it, but we can clearly discern how arguments that are voiced in discussions on theoretical issues in physics are related to one of the two approaches or both.) We shall see in the next section how these different approaches motivate the two different styles of scientific explanations.

4.4. PROBABILISTIC VERSUS TOPOLOGICAL VERSIONS OF THE INDUCTIVE INFERENCE

In the last few sections we have demonstrated how both measure and topological spaces can be presented as specimens of a more

---

[28] It is tempting to say that topological dynamics furnishes us with *de-re* explanations in the sense that these explanations apply to a particular system, while in ergodic theory the explanations are *de-dicto* because they apply to any system that satisfies

general space $\langle S,B,W,T \rangle$. This representation is interesting not only because of its abstract nature, but also because it can help us to illuminate the functions of both the concepts of probability and approximation in statistical physics. Both probabilistic and approximative concepts are used in SM to construct inferences whose abstract forms are the following:

1. We know that $s \in A$.
2. We know that all of the A save a "small set" are B, that is, $\{s : s \in A$ and $s \notin B\} \in W$.
3. We conclude that we can treat s as if it were in B.

The question of the justification of the inference concerns the interpretation of W as a collection of small sets. What makes a set $A \in W$ small is that the possibility that an arbitrarily chosen $s \in S$ is a member of A may be ignored. The definition of a σ-ideal merely establishes closure conditions. Therefore, the members of W must be given a more substantive characterization.

The probabilistic interpretation of W makes it the set of all m-null sets relative to a particular measure m. According to this interpretation, a set B is small if the probability is zero that an arbitrarily chosen $s \in S$ is to be found in B; that is why we can "ignore" this possibility. But what makes a set A an m-null set? The reader should be well aware by now that the answer to this question divides statisticians and philosophers into two camps: On the one hand, there are those who maintain that it is possible to answer the question by citing an objective fact concerning A. On the other hand, there are those for whom m is always merely a subjective measure of ignorance. Some objectivists state that if $\langle S,B,m \rangle$ is the phase space of M, a set A is an m-null set if the average time spent by M in A is zero. This answer would be satisfactory if sets were null given any initial condition of M. Usually, however, the most we can hope for is to be able to say that a set A is null if the average time spent in A is zero relative to a set of initial conditions with a full m-measure. This definition is, of course, blatantly circular, and so it does not meet the subjectivist's challenge

---

the description. This statement, however, should be taken with a grain of salt, because the distinction between *de-re* and *de-dicto* as it applies to physical systems is, itself, a theoretical and not a metaphysical distinction. This point is particularly clear when stationary systems are concerned. By definition, the evolution of a stationary n-particle system M that is in a state S will be the same regardless of any other information about M or the time of observation.

to find a complete noncircular interpretation of probabilities in terms of a nonprobabilistic physical property.

Within the approximative framework, a set is small if it is a set of the first category. That is, sets are small if they are countable unions of sets whose closures have no interior points. The idea behind this characterization is that when a set has no interior points, it is always possible to approximate states belonging to the set by "nearby" states that lie outside the set. In addition, a countable union of such "small" sets is expected to remain small. This characterization, however, leaves the choice of metric unmotivated, and in this case, too, we find two camps: Some maintain that when physical systems are modeled in topological spaces, the metric must be given an objective interpretation, while others doubt that this is possible and conclude that the choice of metric is always idiosyncratic; the operational constraints imposed by the background theory do not single out a unique metric.

Some readers may want to know what type of objective characterization can be given for the choice of metric. Let us elaborate one idea that may clarify this notion. Let $\langle S,d \rangle$ be the phase space of M. The distance $d(s, s')$ between the states s and s' can be taken to be small if, given that M is in s, it can be prepared to be in s' without a large investment of time and energy. This idea can be made precise and is relatively unproblematic when s and s' lie on one and the same trajectory. In such a case, if $d(s, s')$ is small, M will evolve from one of these states to the other quickly. But when M will not evolve "naturally" to s', the physicalist characterization of distance becomes more problematic. We shall return to the discussion on this issue at the end of this chapter.

So far we have concentrated on the similarities between the probabilistic and approximative versions of the inference.[29] Let us not forget, however, that the simultaneous application of the two versions

[29] One of the better known articles in which the different versions of inference are formulated and compared is Ulam and Oxtoby "Measure Preserving Homeomorphisms and Metrical Transitivity," *Ann. Math.* 42 (1941). The authors write: "One may try to make precise the idea that metrical transitivity is the general case by introducing a measure in the space of measure-preserving transformations of E, but it seems difficult to do this in any natural way. Nevertheless, there is a simple and natural metric in the space of automorphisms of E, in case E is compact, and one may ask whether in this space metrical transitivity is the general case in the topological sense, that is, whether such automorphisms consist of all but a set of first category" (p. 877).

of inference could lead to contradictions. Recall that the real line, for example, can be decomposed into a null set A and a set of first category B! Even more generally, the probabilistic and approximative versions of the inference stand for genuinely different types of inferences. From this perspective, it is the similarities and not the differences that are surprising.[30]

The probabilistic version of inference, I_p, is the better known of the two. Indeed, it is easy to understand why knowing that M hardly spends any time in A would be a reason to infer, "all other things being equal," that at a given moment M will not be found in A. The problem with I_p, however, is that, relative to different descriptions of M, different conclusions may follow. This problem is known as the reference class problem. The fact that M can be referred to in different ways sometimes makes it genuinely difficult to apply I_p consistently. This problem does not arise for the approximative version I_a. The fact that A does not have interior points means that the members of A cannot be characterized by a recursive definition that would differentiate them from the "nearby members" lying outside A.[31] But, lurking behind I_a is a much more basic problem. If s is near s' and s is in A, why should we conclude that s' is in A as well?

Only one approach that attempts to answer this question comes to mind, and it is one that is not entirely satisfactory. Because our measurements can locate M only within an interval of values, the argument goes, we can require that our physical theories use only predicates that do not change within such intervals in a discontinuous way. Theories that use only such predicates are called *stable theories*.[32] The principle

---

[30] Oxtoby and Ulam write: "It often happens that what is the general case in the sense of measure is exceptional in the sense of category, and vice versa." Ibid, p. 877. In a most illuminating footnote they demonstrate this point by showing that the law of large numbers is false in the sense of category! Ibid, p. 877.

[31] That is, of course, true if we describe the states by using only the parameters relative to which the system is defined. If we added "external" parameters or predicates, the reference class problem could arise in this context as well. In fact, this is precisely what happened with Oxtoby and Ulam's argument. As we shall see later, relative to the class of homeomorphisms, ergodicity is generic. Relative to the class of flows, ergodicity is not generic. Flows are homeomorphisms that are endowed with an additional "external" differentiable structure.

[32] Some readers may feel that we should talk about a *stable interpretation* of a predicate in a model rather than about a stable predicate. According to this approach, then, stability is a constraint that defines the set of "physically permissible interpretations of a theory." For a discussion on the stability dogma, see Abraham's book (ibid, pp. 4, 186, 234).

that physical theories ought to be stable is sometimes referred to as "the stability dogma." The very inferential import of I_a depends, then, on the acceptance of the stability dogma or similar principles. Only if we make such assumptions can we argue that nearby states cannot differ dramatically with respect to any of their physical parameters.

To sum up, the "proto-inductive" inference, I, can be given two interpretations, I_p and I_a. With regard to each, there is a question of justifying our choice of the σ-ideal of small sets W. In the case of I_p, the question concerns the choice of the measure m, while for I_a there are questions concerning the choice of the metric d. In both cases, though, certain objectivists insist that our choices can be objectively motivated, whereas subjectivists stress the epistemic nature of m as a measure of ignorance and of d as a similarity metric. The fact that the reference class problem does not arise for I_a is one of the forces that motivates the approximative reconstruction of SM. When a property holds for a set of trajectories of the second category, every trajectory can be approximated by a trajectory that has that property. This means that when the property of an individual system is explained using I_a, the explanation can ultimately be connected to the approximative properties of the fine grain dynamic of this system itself; this is not true for explanations using I_p. Probabilistic explanations do not enable us to establish anything with certainty about individual systems, and thus we cannot even claim an approximate knowledge.

It is not sufficiently appreciated that subjectivists with respect to m and subjectivism (or conventionalism) with respect to d can be seen as two different Humean positions. In each case, what is disputed is that there are objective facts by virtue of which the proto-inductive inference can be justified. In each case, subjectivism implies the view that the proto-inductive inference depends on *our* dispositions and habits rather than on the objective dispositions of the physical systems whose behavior we try to predict. In each case, though, we are attributed with having different habits. According to the probabilistic interpretation of Hume, we have a habit of forming *partial beliefs* in the face of ignorance. According to the "approximative" interpretation, we form *similarity judgments* between different states that can be represented as a metric on the state space. The difference between the two approaches is not a trivial one.

Now we are finally in a position to sketch the way in which the topological-dynamical framework may yield a nonprobabilistic reformulation of SM. In doing so, we shall present several new ideas, although the most important element of the presentation will be to highlight an undercurrent in the discussion on the foundations of SM that is not often given full attention. The ideas we shall present deserve such attention because they are some of the oldest, yet most interesting in the literature on the subject.

Once the ergodic hypothesis in its simple erroneous form was universally rejected,[33] there was a need to return to the discussion on first principles and to replace it with a suitable alternative hypothesis. Recall that the original formulation of the ergodic hypothesis (henceforth EH) stated that if a system is left alone, it will eventually assume all of the states in its phase space. If this hypothesis could have been satisfied, it would have enabled us to (1) define SM probabilities as time averages, and (2) calculate time averages using phase averages. However, because the concept of a phase filling curve is simply incoherent,[34] the question becomes whether other hypotheses can replace EH while allowing us to achieve the desiderata (1) and (2). The reader should be well aware of the fact that eventually EH was replaced by the quasi-ergodic hypothesis, or the hypothesis of metric transitivity (henceforth MT). The content of MT is the assumption that, if left alone, almost any trajectory of a mechanical system will intersect any set of positive measures in its phase space. However, before MT, another similar, but nonequivalent replacement for EH was entertained. This older idea was the assumption of regional transitivity (henceforth RT), namely, that if left alone, the system

---

[33] See Rosenthal, "Beweis der Unmoglichkeit ergodischer Gassysteme," *Ann. der Physik* 4 (42) (1913), pp. 796–806, and Plancherel, "Proof of the Impossibility of Ergodic Mechanical Systems," *Ann. der Phys.* 4 (24) (1913), pp. 1061–1063. For an article on the subject, see Brush, "Proof of the Impossibility of Ergodic Systems: The 1913 Papers of Rosenthal and Plancherel," *Transport Theory Statist. Physi.* I (1971). See also the discussion in Chapter 3.

[34] It is actually possible to construct a space-filling curve. However, such curves will not pass through the points only once. These curves were introduced by Peano [see his article in *Math. Ann.* 36 (1890) p. 157]. Recall that the requirement that the system's passage through its phase points be unique is important because it originated from the assumption of mechanical determinism; see the discussion in Chapter 3.

will intersect any *open set* in the phase space. (That is to say, the system will come arbitrarily close to all of the points in the phase space.) RT, of course, is the more natural replacement for EH.[35] In fact, it was only when RT was proven unsatisfactory that the idea of replacing EH with MT suggested itself.[36] Observe that RT does not contain probabilistic concepts; hence, when we recount the rise and the fall of RT, we shall in fact be preoccupied with an attempt to give SM a topological reformulation. This aspect of RT was known at the time.[37] It is also worth mentioning that, although RT was never universally accepted, the ideas surrounding it gave rise to some of the most successful subsequent developments in SM. So we shall end this chapter with a brief introduction to the literature that arises from the discussion of the place of topological concepts in SM.

The basic idea behind RT can be traced back to Maxwell's 1879 article on Boltzmann.[38] Recall that Maxwell believed that the slight perturbations due to the collisions with the walls of the container will suffice to cause the system to jump from one trajectory to another. In this way, he thought, the system will eventually approach all of the

---

[35] Von Neumann argued explicitly that, although the topological concepts are more intuitive and the abstract notion of a measurable set does not have a clear empirical content, one should work within the measure-theoretic framework. He wrote: "In the last analysis, one is always brought to the cardinal question: Does P belong to S or not?; where the set S is merely assumed to be measurable. The opinion is generally prevalent that from the point of view of empiricism such questions are meaningless, for example, when S is everywhere dense, because every measurement is of limited accuracy. The author believes, however, that this attitude must be abandoned." See von Neumann, "Proof of the Quasi-Ergodic Hypothesis," *Proc. Nat. Acad. Sci.* 18 (1932), p. 80. The argument he gave for this view is a fascinating and complicated one.

[36] Brush argues that Boltzmann did not see the point of distinguishing between EH and RT. He claims that only those versed in the work of Baire, Borel, and Lebesgue, which at the time was new and esoteric, could have appreciated the difference. See ibid.

[37] In P. and T. Ehrenfest, *The Conceptual Foundations of Statistical Mechanics*, Cornell Univ. Press (Ithara, NY, 1959) (originally published in 1911 in the *Encyclopaedie der Mathematische Wissenschaften*), the following observation is presented as a criticism of Boltzmann's approach: "If one accepts the ergodic hypothesis, then equation (34), [which is the formula for calculating the duration spent in the various parts of the energy surface,] becomes a purely mechanical theorem independent of any 'probability' considerations" (p. 26). In this context, the Ehrenfests clearly refer to RT and not to EH because, on the page that precedes the above remark, they define ergodic systems as ones for which "The G-path describing their motion is an open curve that covers every part of a multidimensional region densely" (p. 21).

[38] See Maxwell, "On Boltzmann's Theorem on the Average Distribution of Energy in a System of Material Points," *Cam. Phil. Soc. Trans.* v (1879), p. xii.

points in its phase space. This idea can be formulated in the following way: For any two points in the phase space s and s', there will be a point in time such that o_s, the trajectory emanating from s, will occupy a point s'' at that time, and a slight perturbation could sway the system from s'' to s'. This construction, which may explain the term "transitivity," can be given a precise mathematical form with the RT requirement, that is, the requirement that there will be trajectories o_s whose closure fills the whole phase space. Boltzmann was aware of the possibility of dense trajectories as well.[39] Therefore, von Plato's argument that neither Maxwell nor Boltzmann really believed in EH is probably true.[40]

In a survey article written in 1911 by the Ehrenfests, the hypothesis of RT was already clearly articulated. The Ehrenfests distinguished the RT hypothesis from EH and criticized its empirical content. "So far," they wrote, "not even one example is known for which the simple G-path [their term for a mechanical trajectory] approaches arbitrarily close to each point in the corresponding energy surface."[41] To be sure, the Ehrenfests knew about *mathematical* examples satisfying RT. For example, they discussed Lissajous' example, "where the G-path can be visualized as an open geodesic line on the surface of the torus."[42] "It is not difficult," they wrote, "to define purely geometrically a set of curves so that each simple curve of the set approaches arbitrarily closely each of the ∞^3 points inside the torus."[43]

The clear distinction between MT and RT, however, was made much later. Only in 1928, in an article entitled "Structure Analysis of Surface Transformations," did Birkhoff and Smith define MT by saying that "a transformation will be called metrically transitive, if there does not exist a measurable invariant set E such that $0 < m(E) < m(S)$."[44] (The

---

[39] See, for example, Boltzmann, "Ueber die mechanischen Analogien des Zweiten Hauptsatzes der Thermodynamik," *J.F Math.* 100 (1887).

[40] See von Plato, "Boltzmann's Ergodic Hypothesis," *Arch. Hist. Ex. Sc.* (1991).

[41] See ibid, p. 22.

[42] See ibid, p. 89.

[43] See ibid, p. 89. The Lissajous example describes the combined effect of two oscillators. Each separate oscillator describes a circular action. Their combination describes an action on the product of the two circles that is a torus. If the periods of the circles are commensurable, the action on the torus will be periodic; if they are not, the trajectories will wander throughout the torus forever and approach most points.

[44] See Birkhoff and Smith,"Surface Analysis of Surface Transformations," *J. Math. Pu. et App.* (1928), p. 380.

fact that it took so long to distinguish between a trajectory that visits every open set and a trajectory that visits every set of positive measure is probably another product of the obscuring effect of the duality between measure and category.) In 1931, Birkhoff proved that MT, or strong transitivity, as he called the property in his proof, implies a coincidence of the time averages relative to a set of full measure.[45] Once the proof was made public, the possibility of treating RT as a physical hypothesis was more or less forgotten. As we shall see shortly, however, the ideas behind RT resurface in some of the most interesting new developments in the mathematical study of dynamic systems.

The idea of approximating space averages with the time averages relative to a dense trajectory is, as we have said, completely natural. It was established rather early that, if there are dense trajectories, there is a dense set of such trajectories. Therefore, not only can we say that each point in the phase space will be approximated by some point on the dense trajectory, but we can also make the type of inference that follows:

If we know that:
1. there is a dense trajectory,
we can infer that:
2. every trajectory can be approximated by a dense trajectory.

Given the usual understanding of the concept of approximation, this inference seems to provide a good justification for taking *any* property of a dense trajectory as one that holds approximately for the whole population of phase points. Therefore, it is an appealing idea to replace space averages with averages relative to a dense trajectory. This formulation can be seen as an instance of I_a.[46] There is, however, a problem with RT that had already been noticed by the Ehrenfests. There are examples that show that trajectories that differ only slightly may have noticeably different time averages. One such example was

---

[45] Birkhoff, "Proof of the Ergodic Theorem," *Proc. Nat. Acad. Sci.* 17 (1931), pp. 655–660. Interestingly, the same volume contains another article by Birkhoff, where he referred to MT as "strong transitivity" and attributed its original definition to Hopf. See Birkhoff, "Proof of a Recurrence Theorem for Strongly Transitive Systems," ibid, pp. 650–655.

[46] Perhaps one may go even further and say that, if a property cannot be approximated by such a method, it is not a proper physical property.

given by the Ehrenfests in their footnotes.[47] They showed that a periodic trajectory (relative to which frequencies are clearly unrepresentative) can be arbitrarily close to a dense trajectory. What is even worse, there is a dense set of points whose trajectories are periodic![48] (Incidentally, this fact was already known to Poincaré.) This example, then, shows rather clearly that our intuitive ideas concerning approximation do not withstand scrutiny. The conclusion that the Ehrenfests drew from their example was that "one cannot extend the Maxwell–Boltzmann justification to quasi-ergodic systems."[49]

To be sure, the fact that the same system can have a dense set of periodic trajectories *and* a dense set of dense trajectories does not yield a contradiction or a paradox. However, periodic trajectories are usually unstable, and even a slight perturbation will usually disrupt the period and sway the trajectory away from its period. Therefore, the fact that such trajectories form a dense set is puzzling. We certainly do not expect to come across such cases in practice. The clear conclusion, then, is that a dense set is not necessarily a "large" set. (Only sets of the second category that can be presented as uncountable intersections of dense sets are topologically large.) This conclusion, which was new at the time, is rather obvious to us now because we are familiar with the example of the set of rational numbers, which is dense but "small" in the sense of having a Lebesgue measure zero. In fact, the set of periodic trajectories is a null set as well. This fact allowed the resolution of the tension between periodicity and transitivity once the issue of how to ensure the "representativeness" of the time averages calculated relative to a given trajectory was paraphrased in the form of three questions:

1. What is the probability that the trajectory we chose is periodic?
2. What is the probability that the trajectory we chose will have well-defined time averages?

[47] See footnotes 89 and 90 in the Ehrenfest book, ibid, p. 89, where they discuss the Lissajous example that Boltzmann wrote about. This fact was noticed by Birkhoff and Koopman, who wrote in "Recent contributions to the Ergodic Theory," *Proc. Nat. Acad. Sci.* 18 (1932), that: "P. and T. Ehrenfest observed that even if the system is quasi-ergodic in the sense that everywhere dense path curves exist, the mean time of sojourn along the path curve through P in a given region M may vary discontinuously from path to path" (p. 279).
[48] For a proof, see Walters, *Introduction to Ergodic Theory,* Springer-Verlag (New York, 1981), p. 129.
[49] Ibid, p. 91.

3. How probable is it that the time averages will be repeatable? That is, how probable is it that a different choice would give us *exactly* the same result?

The idea behind these questions finally led to the familiar ergodic theoretic approach. However, it is quite interesting that as late as 1922 Birkhoff tried to "save" the insight behind RT when he conjectured that the set of trajectories that are not dense has Lebesgue measure zero.[50] In other words, his idea amounts to a replacement of the third question with:

3*. What is the probability that we chose a trajectory that "approximates" every point in the phase space?

The point is that, even if the topological idea of looking for a dense set of dense trajectories was rejected, it was still possible to ask about the *probability* of obtaining a dense trajectory.[51]

Of course, we should remind ourselves that prior to Birkhoff's proof another attempt was made by von Neumann to reformulate the quasi-ergodic hypothesis.[52] His method was to replace space averages, not with a time average taken relative to a single trajectory, but with a notion of *mean convergence* of the time averages relative to a whole neighborhood of trajectories. Von Neumann argued that, for the purposes of SM, it is sufficient to require that the mean convergence (henceforth MCH) relative to different sets of trajectories will be the same,[53] and Birkhoff accepted this point of view.[54] Note that the concept of mean convergence is even more removed from the original concept of frequentist probability because the time averages relative to the individual trajectories need not be well defined.[55] Von Neumann

---

[50] In fact, if $\langle S, A \rangle$ is a metrizable topological space, $T: S \to S$ is continuous, m is a probability measure on the Borel subsets of S such that $m(A) > 0$ for every nonempty open set, and T is ergodic relative to m, then m-almost every trajectory is dense. See Walters, ibid, p. 132.

[51] See *Acta Math.* 43 (1922), p. 113. Notice that ergodic systems also have periodic trajectories. The equivalence of time and phase averages is up to a set of measure zero.

[52] See von Neumann, "Proof of the Quasi Ergodic Hypothesis," *Proc. Nat. Acad. Sci.* 18 (1932), p. 70.

[53] See von Neumann, "Physical Applications of the Quasi-Ergodic Hypothesis," *Proc. Nat. Acad. Sci.* 18 (1932), p. 263.

[54] See Birthoff, "Recent Contributions . . . ," ibid, p. 281.

[55] In Birkhoff's words: "As Mr. von Neumann pointed out, then, this positive theorem raised at once the important question as to whether or not ordinary time averages exist along the individual path curves" ("Recent Contributions . . . ," ibid, p. 281).

proved that MCH is equivalent to MT.[56] His proof, though, is stronger than Birkhoff's, because it contains some information regarding the rate of convergence. Note that the notion of mean convergence presupposes a notion of distance as well.

4.6. INSTABILITY AND REGIONAL TRANSITIVITY

In the years 1931–1934, there was a flurry of activity by mathematicians who attempted to provide a correct mathematical formulation of the intuitions of the physicists who developed SM.[57] We have already mentioned the work of Birkhoff and von Neumann; to these names we must add Koopman,[58] whose work on operator theory made the subsequent developments possible, as well as Hopf,[59] Hedlund,[60] and Morse. An article entitled "Does Instability Imply Transitivity?,"[61] written by Morse in 1933, has a particular relevance for our discussion because it provides proof that even *after* Birkhoff proved his theorem there was still some interest in the investigation of RT. By examining the ideas presented in this fascinating article, we will have an opportunity to encounter another family of related ideas, namely, ideas about the relations between SM and the concept of instability.

The question raised by Morse in his article involves the most basic issues concerning topology. Recall that when we introduced topological dynamics as a foundational attitude, we characterized it as the insistence on the central role that the dynamics of the individual trajectories should play in mechanical explanations. So far, though, we have concentrated on *global* dynamic properties, such as density, that have the following properties:

1. They are hard to verify in practice.
2. They are defined relative to a *system* of trajectories. In other words,

---

[56] In fact, this is true only in "bounded" spaces.

[57] In *Proc. Nat. Acad. Sci.* 18 (1932) alone there were no less than 14 articles related in one way or another to the ergodic hypothesis.

[58] See, for example, Koopman, "Hamiltonian Systems and Hilbert Spaces," *Proc. Nat. Acad. Sci.* 17 (1931), pp. 315–318.

[59] See, for example, Hopf, "On the Time Average Theorem in Dynamics," *Proc. Nat. Acad. Sci.* 18 (1932), pp. 93–100.

[60] See Hedlund, "On the Metrical Transitivity of the Geodesics on a Surface of Constant Negative Curvature," *Proc. Nat. Acad. Sci.* 20 (1933), pp. 136–139.

[61] In *Proc. Nat. Acad. Sci.* 20 (1933), pp. 46–50.

they reflect the relations between the trajectory and its surroundings rather than being properties of individual trajectories.

Morse, who must have sensed these problems, asked the following questions:

Q_1. Are there any *local* verifiable properties that are characteristic of dense trajectories?

Q_2. If such properties exist, when or under what conditions do they obtain?

If such properties exist, they can further motivate the attempts to replace phase averages with time averages, because the local measurable characterization of the transitive trajectories will make the replacement of phase probabilities with time averages relative to the dense trajectories seem even more like an operationalization of the concept of SM probabilities.

Morse's answer was that there are some families of spaces where density follows from a local property. If Hamiltonian systems can be represented in such spaces, then we shall have a local criterion for transitivity! The family of spaces that Morse talked about is the spaces with a constant negative curvature. Such spaces can be constructed from "pieces" of hyperbolic spaces whose edges are pasted together so that the geodesics will continue in an uninterrupted manner past the pasted edge.[62] (The fact that such pasting is possible requires a proof!) Recall that, in hyperbolic spaces, there are many parallels to a given line that pass through the same point. Therefore, neighboring trajectories "diverge" from each other rapidly. This rapid divergence is due to the swift local change of direction along *each* of the trajectories. This rapid divergence or instability was proven by Morse to imply transitivity.

There is a long history of interest in spaces with negative curvature. Hadamard[63] had already investigated them by the end of the nineteenth century, and at the same time Lyapunov formulated the criterion for instability in such spaces.[64] Poincaré was even aware of the connection between the instability of such spaces and various

---

[62] See Ruelle, *Chance and Chaos*, Princeton Univ. Press (New Jersey, 1991), p. 175.

[63] See Hadamard, "Les surfaces a courbures opposes et leurs lignes geodesique," *J. Mat. Pu. et App.* 4 (1898), pp. 27–73.

[64] See Lyapunov, "Probleme general de la stabilite du mouvement" (1892), translated in *Ann. Math. St.* 17 (1946).

stochastic phenomena to which instability gives rise.[65] He formulated the connection in a dictum that still resonates almost 100 years later. Chance phenomena, Poincaré wrote, can be characterized generally as being the result of processes that have the following property: slight variations in the initial conditions subsequently lead to large variations. Poincaré explained this informally by saying that, when "sensitive" processes are concerned, approximate knowledge of the initial conditions will not have a predictive value because states that are compatible with the approximate descriptions will tend to evolve into an "evenly distributed" set of future states. This feature was characterized as an asymptotic loss of information, which means that "after a sufficiently long time" there will not be any probabilistic dependence between the initial conditions and the subsequent future. Using this characterization, Poincaré argued that even deterministic processes can behave like chance processes, that is, his characterization of chance was independent of questions of determinism. Duhem, too, wrote about the topic. He emphasized the general philosophical importance of the discussion on deterministic but unstable systems.[66]

Morse, though, was the first to move beyond these attractive but vague ideas. He formulated a precise connection between instability and the emergence of stochastic behavior. He proved that instability implies transitivity, and that transitivity may provide a justification for replacing phase averages with time averages. Therefore, Morse demonstrated that instability was a sufficient condition for the replacement of space averages with time averages and that, in effect, instability provides us with a local and verifiable condition the fulfillment of which justifies the use of the microcanonical distributions for predicting the behavior of mechanical systems.

Shortly after Morse, Hedlund[67] and Hopf[68] independently proved that spaces with constant negative curvature are ergodic; that is, they proved that instability implies MT as well as RT. Later they both generalized their theorems to spaces with variable negative

---

[65] See Poincaré, "Chance" in *Science and Method*, Dover (New York, 1957).
[66] See Duhem, *The Aim and Structure of Physical Theory,* Princeton Univ. Press (New Jersey, 1954), p. 138.
[67] See Hedlund, "On the Metrical Transitivity of the Geodesics on Closed Surfaces of Constant Negative Curvature," *Ann. Math.* 35 (1935), p. 787.
[68] See Hopf, "Fuchsian Groups and Ergodic Theory," *Trans. Am. Math. Soc.* 39 (1936).

curvature.[69] These theorems are, in and of themselves, very important, although they left many important questions unanswered:

1. Are spaces with negative curvature even encountered in practice?
2. Can Hamiltonian systems be represented as actions on spaces of negative curvature?

Ulam and Oxtoby proved in 1941 that the ergodic case is "generic" among continuous homeomorphisms.[70] Hence, they argued, we can apply I_a and ignore the nonergodic cases. Later, the validity of their conclusion was challenged when it was discovered that ergodicity was not generic among Hamiltonian flows, a subset of the class of confinuous holomorphisms. (This result is a byproduct of the Kolmogorov–Arnold–Moser theory.[71]) But only in the 1960s, when Sinai actually constructed a model of the ideal gas, which he represented in a space of negative curvature, was the full significance of this line of research widely noticed.[72]

Let us go back to the billiard-ball model of an ideal gas. The model, as you may recall, consists of hard spheres contained in a rigid box. The spheres collide with each other in a perfectly elastic way, obeying the law that the angle of reflection is the same as the angle of impact. Remember that, in the simplified version that we presented in Chapter 2, we had only two spheres, and that one of them was fixed. If the "obstacle" (the fixed sphere) is convex, the distances between nearby trajectories will be amplified, at least for a while, until only one of them will hit the obstacle while the other escapes; from that point on, the motion of the two trajectories will be totally independent. The dynamics of such a system can be represented as a motion along a space with a negative curvature. In such spaces, the motion along each of the trajectories is constantly locally unstable. This property is a verifiable local property of each of the trajectories. As we mentioned earlier, Sinai proved that such a motion is ergodic, which implies that the limiting

---

[69] See Hedlund's abstract 46-3-173 in *Bull. Am. Math. Soc.* and Hopf, "Statistik der geodesischen Linien in Mannigfaltigkeiten negativer Krummung," *Ber. Ver. Sac. Akad. Wiss. Leipzig* 91 (1939).

[70] See ibid.

[71] For a review of the KAM theory, see Walker and Ford, "Amplitude Instability and Ergodic Behavior for Conservative Non-Linear Oscillator Systems," *Phys. Rev. V.* 189 (1) (1969).

[72] See, for example, Sinai, "Dynamical Systems with Elastic Reflections," translated in *Russ. Math. Sur.* 25 (2) (1970), pp. 137–189.

average behavior of almost every trajectory is the same, relative to the natural Lebesgue measure. Hence, the average limiting behavior will be almost always the same, independently of the initial direction of the departure from equilibrium. Consequently, the probabilities we should assign to the events that the sphere will pass through certain regions in the box should be relatively insensitive to additional information about the initial condition of the system as long as it is not infinitely precise. However, even if we had knowledge of the exact state of the system that would allow us to say with certainty how it will evolve, the probabilistic assignments would still be meaningful. They would rely on the property that the motion is locally unstable everywhere, a property that remains meaningful even for an omniscient observer. For such an observer, the fact that the probabilities are related to a local property of the individual trajectories supplies them with empirical content that remains meaningful even if the probabilities are no longer needed for the purposes of prediction. In such cases, instability becomes the "ghost" of the departed probabilities, so to speak.

An important byproduct of this account is that it also supplies us with an explanation as to why ignorance with respect to the exact (or even the approximate) state of an unstable system is to be expected. The existence of singularities (e.g., because of the collisions) makes it impossible to calculate the state of such systems, and the nonlinearity of the motion makes it difficult to employ approximation techniques. Therefore, instability functions both as an integrating concept and a physical reason for the need to use probabilistic reasoning. It is desirable that integrated conceptions of probabilities in other contexts have this double function as well.

4.7. CAN SM BE FORMULATED USING ONLY TOPOLOGICAL CONCEPTS?

This chapter centered around an attempt to provide a new condition, the satisfaction of which justifies the replacement of phase averages with time averages; this condition is RT. The new rationale it provides for replacing the phase averages with time averages is the following: RT guarantees that the "generic" trajectory is dense. Now, when the time averages are taken relative to a dense trajectory every point in the phase space is approximated by some point on the dense trajectories. Hence, the argument concludes, the time and space averages must be interchangeable. This rationale is close in spirit to the naturalist's conception of physics, which emphasizes the explanatory significance of the dynamic properties of the individual trajectories. RT, you may recall, depends on the fine grain details of the dynamics of the individual trajectories. Indeed, if we could have required RT, instead of MT, we would have had a probability-free version of SM that is harmonious with the naturalist's approach. We saw, however, that the use of some probabilistic concepts is unavoidable for the formulation of SM. Therefore, the naturalist SM that we described is a nonstarter. Nevertheless, a mixed approach to SM, where we define local dynamic properties such as instability and prove that m-almost every trajectory has these properties, was proven to be a very powerful approach. Moreover, it is our contention that this mixed conception lies behind Poincaré's definition of chance. We shall continue to discuss this approach in the next chapter. However, let us end this chapter with a few more general remarks concerning the relations between topological and probabilistic concepts.

The idea that the choice of probabilities in the context of SM must be constrained seems to be essential for an understanding of SM probabilities. The formulation of such constraints usually necessitates the introduction of a topology or a metric on the points of the phase space, as well as a presentation of some "appropriateness" conditions that constrain the simultaneous choice of a measure m and a metric d. One of these constraints is regularity, and another is the requirement that $m(A) > 0$ for every open set A. A related but different constraint is continuity: If A_ε is a small translation of A, $m(A)$ should equal $m(A_\varepsilon)$. These constraints are all equivalent to limiting our choice to measures that are absolutely continuous with respect to the Lebesgue measure.

The constraints that we have mentioned can be given "physicalist" interpretations. The idea that open sets have positive measures stems from the facts that open sets are "stable."[73] This means that a slight perturbation will leave a member of A inside A. (This stability justifies the idea that open sets are the denotations of the ranges of the possible results of well-designed experiments). In turn, this quality explains

[73] Malament and Zabell, in "Why Gibbs Phase-Averages Work?," *Phil. Sci.* 47 (1980), attribute this idea to Lebowitz and Penrose in their "Modern Ergodic Theory" (*Phys. Today*, 1973) They fail, however, to acknowledge the existence of a topological concept of "genericity" that has its own raison d'etre although it might conflict some of the time with the measure-theoretic concept of a set of full measure. Hence, they do not take this idea very seriously.

why we should regard open sets as nonnegligible. Regularity can be motivated by the idea that the measure of a set is observable and hence we should be able to approximate it using open sets. The continuity assumption can be motivated by the observation that the small translation A_ε can be thought of as a small perturbation of all of the elements of A.[74] Therefore, the two sets have the same members and should be given the same measure. It is crucially important, though, to realize that these "physicalist" interpretations of the constraints are not much more than heuristics. As they stand, these interpretations introduce concepts that are sensitive to the choice of the perturbing mechanism. This sensitivity, which is no trivial matter, renders both constraints useless from a foundational point of view.

Of course, it is possible to introduce regularity and continuity as abstract constraints. Jaynes,[75] for example, and Malament and Zabell who followed him, introduced continuity as an abstract symmetry relation. However, if we regard the topological constraints as abstract symmetry relations, they cease to be of use to the objectivist for whom these constraints are supposed to function as *justifications* for his choice of probabilities. If continuity and regularity are merely abstract symmetry relations, what could justify the idea of imposing them as constraints?

There are many other connections between topological and probabilistic concepts that can function as constraints on the simultaneous choice of both. For example, there are many cases in which transitivity implies ergodicity.[76] In such cases, the choice of a metric relative to which a transformation is topologically transitive may motivate the choice of measure relative to which the transformation is ergodic. Of particular importance are cases in which it is possible to decompose the phase space into topologically invariant tori, as is the case with the Anosov systems.[77] When such systems are concerned, regularity implies ergodicity within the invariant torus! Another fascinating connection is the relation between ergodicity and topological entropy. This allows us to characterize ergodic measures as those regular measures that satisfy a certain operational constraint formulated in purely topologi-

[74] This is the best case we could make for the argument of Malament and Zabell, ibid, p. 346.
[75] See Jaynes, "A Well-Posed Problem," *Foun. Phys.* 3 (1973), pp. 477–492.
[76] Affine 1-1 transformations are topologically transitive iff they are ergodic. See Walters, ibid, p. 132.
[77] For a discussion on Anosov systems, see, for example, Mane, ibid, pp. 166–207.

cal terms.[78] This characterization narrows the choice up to equivalence. All of these connections are the result of the basic duality between probability and approximation and are extremely important to the foundations of SM. These connections point to the usefulness of studying probabilistic and topological concepts together, and of devising rules for the simultaneous choice of probabilistic and metric concepts; they do not imply a reduction of the probabilities or a nonprobabilistic reconstruction of SM.

There is, however, one way in which the introduction of topological concepts may help us to understand the sense in which SM probabilities have an objective character. In those cases where instability supplies us with local verifiable conditions for ergodicity, a condition whose satisfaction depends on the dispositions of the individual trajectories, ergodicity will remain a meaningful property even for an omniscient observer who knows the exact state of the world. This sense of objectivity, though, should be distinguished from others, because it merely points to the possibility that probabilistic concepts acquire a new status when they are deeply integrated within a physical theory.

4.8. ON THE PHYSICAL INTERPRETATION OF THE NOTION OF DISTANCE

Before ending this chapter we would like to make another point. Throughout this chapter we attempted to systematize the idea that probabilistic concepts may be replaced by topological ones. We entertained the possibility that after the replacement we will have a theory that assumes only the notion of distance between points or, equivalently, the notion of an approximating neighborhood. What motivated this attempt was the fact that probabilities cannot be given a "naturalist" or physicalist interpretation. That is why we tried to start afresh, dispensing with probabilistic concepts altogether. We would like to end this chapter with an argument proving that, even if such a conception were possible, the result would be problematic for two reasons. First, as we remarked earlier, the notion of limit behavior transcends the purely naturalistic analysis. Second, as we shall prove shortly, the instrumentalization of the notion of distance necessitates an importation of probabilistic notions. This point will establish new support for

---

[78] See, for example, Mane, ibid, pp. 236–263.

our conviction that both probabilistic and approximative concepts are necessary and that we should not attempt to reduce either one to the other.

Let s, s' be two states that lie on the same phase trajectory. We have seen earlier that, in this case, it is relatively easy to define the physical distance between the two points. The distance $d(s, s')$ may be identified with the time that it takes M to move from s to s' or, conversely, the time that it takes to move from s' to s. This way of measuring distance is appealing; according to this conception, distance is a measure of the difficulty of transforming a system that was prepared to be in s into a system that occupies the state s'. That is, distance may be regarded as a disposition. This method works well in a limited number of degenerate cases. (It applies, for example, to systems with a single degree of freedom. As you may recall, such a system has a single trajectory that passes through all of its points.) A general definition of distance, though, requires that we define distance in two other cases as well:

1. when s' is a limit point of the trajectory of s;
2. when s and s' lie on different trajectories.

Let us consider the first case. Ordinarily, we would define the notion of a limit point in the following way. s' is a limit point of $O_T(s)$ iff there is a subsequence $n_i \uparrow \infty$ such that $d(s', T^{n_i}(s)) \rightarrow 0$. This case applies when we describe the motion of certain dissipative systems. Our damped oscillator is an example of such a system. (As you may recall, a pendulum that swings to a halt is a damped oscillator.) Now, observe that, when s' is a limit point of $O_T(s)$, then

(1) $s' = T(s')$.

That is, s' is a fixed point. Second,

(2) for every n, $T^n(s) \neq s'$.[79]

Therefore, if $d(s, s')$ is the time that it takes M to move from s to s', then $d(s, s') = \infty$. This definition, though, is not satisfactory. We need a distinction between the case where s' is a limit point and a case where it is not. [An example of this is when $O_T(s)$ approaches a limit cycle.]

[79] Otherwise, there will be multiple trajectories emanating backwards from s'. The first is the "fixed point trajectory," s', and the second is $O_T(s)$. This is contrary to the uniqueness of the trajectories that emanate from every point.

However, to articulate the difference between the two cases we need a prior notion of distance that allows us to say that, in the second case, there is no subsequence $n_i \uparrow \infty$ s.t. $d(s', T^{n_i}(s)) \to 0$.

The problems become even more difficult when we attempt to define the distance between points that lie on different trajectories. In such a case, even if there is only a small variation in all of the values of the physical parameters that define s and s', there may not be any obvious operationalizable meaning to the expression that the distance between s and s' is small. There may not be any simple procedure for transforming a system that was prepared to be in s into a system that occupies the state s'. How, then, shall we define the notion of distance for arbitrary pairs of points? One idea in this direction is the following. Given that M was prepared to be in s, we may ask ourselves how much energy it takes to perturb M so that it occupies the state s'', a state that is on the trajectory of s', and then measure the time that it takes M to move to s'. Therefore, according to this idea, the distance between s and s' is a quantity that depends on two factors, namely, on the perturbation energy and on the time. This idea is attractive because it seems to follow the logic of the measurement of distance in the simple case. Here, too, we seem to be measuring how difficult it is to transform M to occupy s' after it was prepared to be in s; only this time we add to our vocabulary a perturbing mechanism.[80]

Upon reflection, though, it becomes clear that the concept of distance that we are considering raises some serious objections. First, one should note that, when many-particle systems are concerned, we are simply incapable of controlling the effect of a perturbation on a given system to an arbitrary degree of accuracy. Next, observe that perturbations change the energy of the system. Strictly speaking, this means that s' does not belong to the phase space of the perturbed system. (More precisely, s' is on a different energy hypersurface.) For the same reason, the perturbed trajectory does not belong to the original phase space either. This fact creates new difficulties. First, there are problems concerning the identity of states across phase spaces. Second, we need an identity criterion for trajectories across phase spaces. Finally, we have to confront a more radical objection. The

---

[80] There may be various ways to perturb M from s to a state that lies on the trajectory of s'. Relative to each, there will be a different perturbation energy and different time of sojourn to s'. Consequently, we may either have to devise an energy–time measure, accept incommensurable cases, or average the results.

attraction of the concept that we are discussing stems from the fact that it allows us to think of distance as a disposition of a particular system. If, however, the perturbation destroys the identity of the system, then the behavior of the system after the perturbation is not necessarily a disposition of the system that existed even before the perturbation. Therefore, invoking perturbations will not help us in our quest for a naturalist or a physicalist definition of distance.

It is quite interesting from a philosophical point of view that the question of the identity criteria across phase spaces received a lot of attention from mathematicians and mathematical physicists. In fact, there is a theory, with a considerable philosophical import, where we can find criteria that may allow us to decide whether a perturbation destroys the identity of the system or not. The idea is as follows. Fix a phase space S, devise a perturbation method, and apply the same perturbation to all of the trajectories of the original phase space.[81] As a result, you will obtain a new phase space S'. In some cases, S and S' will have the same structure, which means that the "qualitative" limit behavior of the trajectories of S will be the same as the limit behavior in S'. For every limit point in S there is a limit point in S', for every limit cycle in S there is a limit cycle in S', for every limit torus in S there is a limit torus in S', and so on. No new limit sets are created and no limit sets are destroyed. Systems with this property are called *structurally stable systems*. The idea, then, is that the identity of a structurally stable system may be regarded as being sufficiently resilient so as to survive a slight perturbation. The identity of systems that are not structurally stable, on the other hand, may be "destroyed" by even a slight perturbation. In mathematical terms, this means that the topological character of the phase space of structurally unstable systems is not preserved under a perturbation. On the other hand, when structurally stable systems are concerned, slight perturbations will not change the character of the space. Therefore, it is possible to choose some canonical way of assigning to each trajectory of S a trajectory of S'. To each point s in S we may assign a neighborhood of points in S', the members of which may be regarded as the "counterparts" of s in S'. This idea guided Kolmogorov, Arnold, and Moser when they proved that Hamiltonian systems are structurally stable. They

[81] This means that we parametrize each of the trajectories, fix a state where $t = 0$ on each trajectory and apply the perturbation to all of the trajectories.

demonstrated that slight perturbations will not cause a conditionally periodic system to lose its orbit completely. (For example, the gravitational field created by a comet will not destroy the planetary system.) This remarkable result, then, proves that n-particle systems may be attributed with a continuous identity, even when it is perturbed. So, finally, we have an argument proving that the inclusion of a perturbing mechanism in the definition of phase space distance does not, in and of itself, invalidate the definition.

The first objection, though, still stands. Many-particle systems may be structurally stable, but they are not stable in the sense of Lagrange. This means that two trajectories, which at one point in time cannot be told apart, may diverge very rapidly thereafter. The implication of Lagrange instability is that it is impossible to control the precise effect of a perturbation; slight imprecisions may cause large deviations. Therefore, the effect of a perturbation has to be described as the expectation value of the various possible precise effects it might have. More precisely, to measure the distance between s and s' using the effects of the perturbation we need to determine, for each of the possible states s'' (s'' is a state that M may assume after the perturbation) whether $O_T(s'')$ includes $s*$, a member of $S*$ ($S*$ is the set of the counterparts of s') and then measure the time that it takes to move from s'' to $s*$. The effect of the perturbation, then, may be identified with the weighted average or the expectation of the various possibilities. Observe that this definition refers to a probabilistic concept, namely, to the measure relative to which we calculate the expectation value. Therefore, even if the idea of using perturbations to extend the notion of distance does not lead to a circular argument, it refers to probabilistic notions, and, hence, all of the problems regarding the physicalist interpretation of probabilities are imported into the discussion on the concept of distance.

Before we continue we should mention an issue that will resurface later. It concerns the status of the Lebesgue measure. This measure, as you may recall, is defined as an extension of the concept of a length of a line segment or the volume of a cube. Now, after our discussion on the concept of distance, we may explain why the Lebesgue measure does not solve the problems with the physicalist interpretation of probability. Let us begin by reviewing a particularly simple way to construct the Lebesgue measure of a set A. The first step is to define the notion of a measure for the open sets. (One way of doing so is to define the notion of a measure for elementary open sets, for example, defining the measure of a line segment as the distance between the end points and extending the definition to an arbitrary open set.) The next step is to define the measure for an arbitrary measurable set A. We may do so by setting

$$m(A) = \inf\{m(B)|B \supseteq A \text{ and } B \text{ is open}\}.$$

In general, it is desirable that we have in addition

$$m(A) = \sup\{m(D)|D \subset A \text{ and } D \text{ is compact}\}.$$

So, those sets A for which

$$\inf\{m(B)|B| A \text{ and } B \text{ is open}\} =$$
$$\sup\{m(D)|DC A \text{ and } D \text{ is compact}\}$$

have a well-defined Lebesgue measure.

In many textbooks on mathematical physics it is assumed that the phase space has a "built-in" natural Lebesgue measure that determines the notion of phase volume. When such an assumption is made, we may define the notion of integration in the phase space by using the Lebesgue measure and stipulate that the phase probabilities have to be absolutely continuous with respect to the Lebesgue measure. (This means that zero Lebesgue measure implies zero probability.) In some cases, this assumption defines the probabilities uniquely (for example, when the underlying system is ergodic). Therefore, many physicists assume that physical probabilities have a straightforward meaning. This view, we believe, is mistaken. The problem with the use of the Lebesgue measure to define probabilities is that it relies on a prior definition of the distance between arbitrary points. As we tried to argue, such a definition is either circular or it depends on prior probabilistic notions. Our conclusion, then, is that the Lebesgue measure, as important as it is, does not solve the foundational puzzle with which we are concerned. When we use the Lebesgue measure to define probabilities, it needs proper foundations itself.

189

5

Three Solutions

5.1. THE PUZZLE CONCERNING THE CONCEPT OF PROBABILITY IN SM

Throughout the book we have been preoccupied with a single simple scenario. We consider now an n-particle system M that is in a state s_t at t. We want to know whether s_t has a certain property A. The question that we are asking has a perfectly clear physical meaning. Nevertheless, the information that the answer requires may not always be available to us. It may be the case that the experiment needed to answer the question is too expensive or difficult to perform. It is also possible that there are theoretical reasons why we cannot perform the experiment, or, simply, the experiment has not been performed yet. In all of these cases the need for probabilities arises. We need to calculate the probability $P(s_t \in A)$ and use our results to derive predictions concerning M. But, even if there is little doubt that we constantly use probabilities, there is a basic problem concerning the interpretation of the probabilities. It is difficult to define probabilities as observable parameters that depend only on the physical properties of M. This difficulty gives rise to a philosophical puzzle: If probabilities are not physical parameters that we discover by methods of observation and measurement only, how can we justify our willingness to be guided by them? How can we explain the utility of probabilities? In the previous chapters, we reviewed several attempts to solve this problem. In each chapter we singled out an idea that seemed promising, we gave it a charitable exposition, and then we subjected it to a critique. Now we shall attempt to go one step further. We shall offer an outline of three different solutions to the problem. The solutions are obtained by learning from the shortcomings of the earlier attempts discussed thus far.

The first solution we will consider is subjectivist in spirit. We shall attempt to formulate a new subjectivist conception that is consistent

190

with de Finetti's ideas, but clearly broader, and that may allow us to present a new version of SM. This version, we believe, is less problematic than other attempts to give SM a subjectivist interpretation. The second solution we shall discuss involves a reformulation of the ergodic approach. This approach, which is clearly the origin of some of the most important ideas about statistical mechanics, so far has proven to be rather difficult to present as a foundational approach. In this chapter we shall attempt to rethink what is distinct and important in this body of ideas. Finally, we shall return to Gibbs's pragmatist approach. This approach gave rise to a whole slew of problems. In fact, this book would not have been written if Gibbs's approach were less problematic. However, in the final analysis, we believe that Gibbs's approach, complete with its inseparable philosophical underpinning, is a most convincing account of SM.

One of the outcomes of our earlier discussions was the development of a conviction that questions concerning probabilities should be studied together with questions concerning the concept of approximation. Approximations, like probabilities, are tools that we use to derive predictions when we lack complete and precise knowledge. And in this context, too, the question arises: Why are we willing to trust approximate reasoning? Is there a physical reason why approximations "work"? If s^* is a state that approximates s_t, and s^* has a certain property, what may we legitimately conclude about s_t? More generally, what does it mean to say that two states are near each other? Is the distance between states a physical parameter? These questions are, obviously, analogous to those that we asked about the nature of probabilities. But, as we argued in Chapter 4, the analogy runs deep; in fact, the two sets of questions are profoundly related. (Mathematically, their connection is expressed in the duality between topological and measure-theoretic concepts. Philosophically, these concepts lay behind two different styles of inductive explanations.) Consequently, we shall attempt, whenever possible, to address the two sets of problems together.

It should be emphasized again that the issues that concern us are not simply technical ones. If doubts concerning the adequacy of the definitions of probability and approximation, as they appear in physics, are substantial, they might cast a wide and dark shadow on physics as a whole. We have had plenty of opportunities to see how probabilistic and topological concepts are used to define a variety of important physical concepts. Indeed, the very concept of an observable physical quantity presupposes the topological notion of an approximating neighborhood. The concept of measurement necessitates the introduction of integration over time, a concept that is defined using topological and probabilistic notions. And the idea of expressing a function as an infinite linear combination of the elements of a basis, the very foundation of Hilbert space theories, relies on a notion of convergence that uses measure and topology. These concepts, in turn, are used to define an even larger set of physical concepts. So, finally, if there are serious doubts concerning the physical meaning of convergence and integration, these doubts will certainly affect most of the areas of physics. That is, if there are problems concerning the definition of probabilistic and topological concepts, these problems will be inherited by many other concepts.

5.2. THE EXTENSION OF THE SUBJECTIVIST FRAMEWORK

The idea that probabilities should be interpreted as degrees of belief unites subjectivists of all shades and stripes. Subjectivists believe that objective probabilities are fictitious entities whose postulation obscures the role that probabilities play in science. For some subjectivists, the reinterpretation of probabilities is the sole content of the subjectivist stance. But, behind the preoccupation with probabilities, it is possible to discern a more general philosophical credo. Subjectivists maintain that our scientific concepts, in general, should be presented in epistemic terms as propositional attitudes. This mode of defining concepts is advantageous because it highlights their function in scientific inquiry. On the other hand, the reification of our concepts, that is, their presentation in terms that refer to independently existing entities, is considered by the subjectivist to be, simply, bad epistemology; it adds metaphysical mysteries, rather than solving the epistemic ones. For the subjectivist to talk about objective probabilities is a mode of expression as objectionable as, say, expressing one's hope that A will happen by using the sentence "A is objectively hope-worthy." The postulation of such an objective property does not take us any closer toward an understanding of hope. It merely complicates things by presenting us with a new problem, namely, we have to wonder how the comprehension of objective hope-worthiness is possible.[1]

---

[1] There are interesting similarities between subjectivism and constructivism in mathematics. In both cases, there is an emphasis on describing the function of our concepts and a resistance to excessive reification.

The general aim of the subjectivist epistemologist, then, is to describe the process of inquiry. The most important question, from this point of view, is what is involved in the acquisition of a new belief. The subjectivist usually gives this question a "dynamic" answer. According to the subjectivist, we should be preoccupied, first and foremost, with the effects of new information on our current body of beliefs. The acquisition of new knowledge will induce us to change our degrees of belief. This change, in turn, will have an effect on our dispositions to accept or reject new hypotheses. According to the subjectivist, then, the task of the epistemologist is to describe this dynamics.

At this point probabilities enter the picture. For the subjectivist, the dynamics of inquiry is governed, essentially, by the following rule: If, prior to your knowing that B, your conditional probability $P(A,B)$, equals r, then, other thing being equal, your unconditional probability $P^*(A)$, which you form after finding out that B should equal r as well. This rule, as you know, does not constrain the probabilities uniquely. More generally, the subjectivist tolerates the fact that different agents might have different degrees of belief. The only requirement that is imposed on the system of degrees of belief is internal coherence. Therefore, according to the subjectivist, there is an ineliminable subjective element in every inquiry.

Notice that this "single inquirer" perspective of subjectivism places it in strong opposition to the Fregean tradition. Frege presented an account of meaning that centered on communication. Therefore, he tended to postulate independently existing entities. (It is easier to explain communication when external objects are presumed present.) De Finetti's epistemology, on the other hand, centers around the changes in the dispositions of a single inquirer. In such a context communication is not a central issue,[2] and therefore there is no temptation to postulate independently existing entities. Consequently, de Finetti developed a single-inquirer perspective that discouraged reification and encouraged a functional account of meaning. This form of subjectivism, then, should not be construed as an attempt to deemphasize cognition. Neither is it an attempt to celebrate the uniqueness of one's own point of view. It is, primarily, a functional and dynamic

---

[2] Issues concerning intersubjective agreement are addressed by characterizing special scenarios where the probabilities of different agents converge. Such may be the case, for example, when different agents, whose initial probabilities are not mutually singular, change their probabilities in response to the same information.

perspective that tends to emphasize epistemic issues and, secondarily, a critique of communication-centered epistemologies that tend to reify our concepts.

With these general remarks in mind, we may proceed to describe a subjectivist perspective on various concepts that do not involve probabilistic reasoning. For example, we may attempt to outline a subjectivist account of approximation judgments. Later, we will combine both the subjectivist account of probability and the new theory of subjective topology and attempt to construct a more comprehensive theory in terms of which we will explain some of the puzzles that arise in statistical mechanics. It should be noted that the general subjectivist methodology that we outlined may be applied to other concepts as well.

Let M be, as before, an n-particle system. Let s and s' be two possible states of M. We suppose that the states s and s' are such that the positions and the momenta of all of the particles are only slightly different in the two cases. This, we argued, does not always imply that M may be transformed from s to s' in an easy and fast way.[3] How, then, shall we understand the notion of distance? The subjectivist answers that, when we say that s is close to s', what we mean is that s is similar to s'. Why, though, do we form similarity judgments? What is the function of judgments concerning the distance between s and s'? The subjectivist answers in the following way. We are called on to imagine two hypothetical epistemic situations E and E'. In the first, we are attributed with the same body of beliefs that we currently hold, but we also learn that M is in a state s. In the second hypothetical epistemic state we find out, instead, that M is in the state s'. Having identified E and E', we then begin reflecting on the differences between the two epistemic states. We do so by noting the consequences of the hypothetical additions to our current body of beliefs. Once this is done, we may form a judgment concerning the degree of similarity between E and E'. Assessments of this kind can be a basis for a general notion of degree of similarity between possible epistemic states. When such a notion is defined, we may measure the distance of s and s' in terms of the degree of similarity of E and E'. You may notice how this account translates a question that is, seemingly, about an objective relation between states to an epistemic one that centers on the function of the relation in scientific inquiries. The assessment of the epistemic difference between E and E' allows us to form an opinion concerning the

---

[3] See the end of Chapter 4.

urgency of deciding between them, that is, deciding on the basis of our current knowledge whether M is in state s or s'. If s and s' are similar, the potential error that we commit by confusing the two is small. More generally, the notion of similarity should be understood in terms of potential errors.

The subjectivist's interpretation of the concept of distance makes it easy to explain why judgments of distance obey the first two metric axioms. The difference between E and itself is clearly nil, and the difference between E and E' is the same as the difference between E' and E. The third axiom, the triangle inequality, may be regarded as a weak form of transitivity. If E and E' are very similar and E' is very similar to E'', it cannot be the case that E is very dissimilar to E''.

Our next topic concerns the way in which similarity judgments allow us to infer new beliefs from old ones. As we remarked in the previous chapter, similarity judgments give rise to a new form of inductive inference. One way of stating the form of inference is the following continuity principle:

(CP) If $f_A(s) = r$
and $d(s, s') < \varepsilon$,
then, $r - \delta < f_A(s') < r + \delta$.

(f_A is a function that assigns A-values to the different states.) If δ depends only on ε and not on s, we may talk about uniform continuity.

Is it irrational to hold degrees of similarity that do not obey the above requirement? For the time being we do not have a very good answer. A good argument proving the irrationality of incoherent degrees of similarity should be analogous to the proof that incoherent degrees of belief yield a sure-fire loss. We intend to try and find such an argument in the future.[4,5]

Another issue that we need to address is describing the manner in

---

[4] Such an argument is rather easy in those cases where each property defines a separate "dimension of similarity" and the "global" distance is calculated, using the Pythagorean theorem, as the square root of the sum of the squares of the distances. In such a case, a violation of (CH) will result in immediate inconsistencies.

[5] Another important issue is the definition of the relation of equivalence between similarity judgments. Such a definition should play the role that absolute continuity plays with respect to probability judgments. One suggestion is that d and d' are to be considered equivalent if relative to the topology they give in to, the same subsets are considered open. These conditions, incidentally, imply that d and d' also agree on the set of nowhere dense sets.

195

which we ought to revise our similarity judgments in response to new information. One idea is the following. When both s and s' can be located in a single "small" set, their distance must be considered small. The notion of a size of a set A may be given various interpretations. It may be defined simply as the probability of A. We may also define the size of A as its diameter; that is, we may identify it with dim $(A) = \max\{d(s, s')|s, s' \in A\}$. (Clearly, if both s and s' are in A, their distance has to be less than or equal to the diameter of A.)

Other important information that we may employ when we determine the distance between states is the amount of time and energy it takes to transform M from state s into s'. If the two states are easily intertransformable, they should be regarded as close. More precisely, when we are concerned with the distance between states that lie on the same trajectory, we may assign their "trajectory distance" as their degree of similarity. (This idea is consistent with the subjectivist's conception of distance. If you know, for example, that s and s' are easily intertransformable, the corresponding E and E' should be considered similar, because it most likely will be difficult to tell them apart in practice.)

Suppose, then, that we follow the subjectivist's way of thinking; that is, we interpret our probabilities in terms of degrees of belief and our notion of distance in terms of degrees of similarity. This way of proceeding may help us to avoid various conceptual problems. We have seen how the subjectivist framework allows us to distinguish between the relative frequencies, for example, from the probabilities. The former are objective properties of sequences of events, and the latter are subjective. In the case of distance, we may distinguish between the distances on a trajectory and the distance between states that lie on different trajectories. Again, the former are objective properties and the latter reflect subjective judgments. In both cases there is no temptation to resort to circular arguments. But note that, even if we are satisfied by these accounts, when considered separately, we still have to answer a further question, namely, how to justify, on subjectivist grounds, the requirement that our probabilistic and metric concepts should cohere with one another. We already encountered one such connection. The Lebesgue measure has to satisfy the following regularity requirement:

$$m(A) = \inf\{m(B)|\ B \supseteq A \text{ and } B \text{ is open}\} =$$
$$= \sup\{m(D)|D \subset A \text{ and } D \text{ is compact}\}.$$

How, though, can we justify such a requirement in the eyes of a sub-jectivist?[6] Are we not violating the dictum that the only universally binding constraints are those whose violations result in incoherence? Our answer is contained in the following observation. When constraints on our degrees of belief are supposed to be derived from objectivist principles, they may meet with disapproval from subjectivists. The same may be true for the constraints on our degrees of similarity. The principles we are considering, though, are not supposed to be justified on objectivist grounds. When we conceive of our project as a si-multaneous choice of degrees of belief and degrees of similarity, we may view regularity requirements as extended coherence re-quirements. After all, even subjectivists may agree that when we select degrees of belief and degrees of similarity together, we should do it in a coordinated manner. Such coordination is needed to ensure that we do not have conflicting notions of size. This, then, may be a motivation, in the subjectivist spirit, of the regularity principle. It is rather remark-able that the acceptance of the regularity principle is sufficient to deter-mine our degrees of belief up to absolute continuity.[7]

Let us now try to understand what kind of conception of SM is obtained when we adopt the double subjectivist view. As usual, the term subjectivism may be somewhat alarming to some, but we shall attempt to ignore this point. The subjectivist view deserves to be judged solely on its merits.

Notice that, on the face of it, the subjectivist's origins of the con-cepts of probability and distance need not worry us too much. We may regard the foundational activity that led to their definitions as a ladder that may be discarded once it has served its purpose. This, however, is not really the case. From a subjectivist's point of view, the case of SM is a particularly difficult one. We have seen, again and again, that in the context of SM we do not expect to "learn from experience." That is why

---

[6] If the σ-algebra of the measurable sets is defined as the Borel σ-algebra of the open sets, that is, if we generate the measurable events from the open neighborhoods of a set of events, then it is possible to prove that the regularity property is automati-cally satisfied. Every Borel measure on a metric space is regular (see Walters, ibid, p. 147). Therefore, the regularity property presents a nontrivial issue for the sub-jectivists only if they attemp to "harmonize" previously defined notions of distance and probability.

[7] In Appendix I is a proof of a theorem stating that in compact metric spaces regular finitely additive measures are also σ-additive. Therefore, even the issue of the justification of the σ-additivity requirement is resolved when we assume compact-ness on independent grounds.

we use ergodic measures that frustrate any attempt to revise our probabilities in the light of new observations on the underlying system. To be sure, from a subjectivist's point of view there is nothing strictly irrational about this choice. (De Finetti himself talks about ergodicity, which he regards as an extreme possibility.[8]) Nevertheless, our willingness to assume, in the context of SM, that nothing might be gained by observation seems to go against the grain of the inductive method. This presents a serious puzzle to the subjectivist. The only way to escape it is to admit an objective reason, inherent randomness for example, that will explain why the learning process will be frustrated. But this option is not available to the subjectivist. The question, then, is whether the doubly subjectivist view offers a way out. We shall begin by explaining why the doubly subjectivist view is better suited to handle this puzzle than the other versions of the subjectivist view that we presented earlier. Ultimately, though, we are not entirely satisfied with the subjectivist solution.

When the distance between the members of S is defined subjectively, the notion of Lagrange instability acquires a new meaning. Recall that Lagrange instability implies the following property. There are cases where $d(s, s') < \varepsilon$, but $d(T^n(s), T^n(s')) \gg \varepsilon$ even for relatively small n. Stated in subjectivist terms, this property has the following interpretation. Let M be in either state s or in s', where s and s' are states between which we cannot differentiate. (For example, our best measuring devices cannot answer the question of whether M is, in fact, in s or in s'.) Consequently, when we construct the corresponding epistemic states E and E', we do not have any grounds for deciding at the moment whether E should be preferred to E', or vice versa. When M is Lagrange-unstable, though, we know in advance that in the future E and E' may entail entirely different predictions. Therefore, while at present we do not have any grounds for choosing between E and E', we already have evidence that such a choice is, indeed, called for. Now, if such cases were rare or isolated, there would be no reason to be seriously concerned. But, when Lagrange-unstable systems are concerned, we know that the probability of encountering an unstable state is very high. In such cases, then, a problem arises. Our standards of precision are too coarse to allow us to make all of the requisite discriminations. Observe that one of the ways in which to react to an unstable situation is by refus-

[8] See de Finetti, *Theory of Probability*, vol. II, Wiley (New York, 1973), pp. 149–150.

ing to choose between E and E'. In fact, if ε is the resolution of our measurement apparatus, we may refuse to choose between the hypotheses that M is in any state in an ε-neighborhood of s. This refusal may result in a loss of information. However, instability means that states from a relatively small neighborhood will eventually "spread out" all over S. Therefore, we may conclude that the information was not very valuable to begin with. There is, in fact, a theorem that makes this intuition precise. As we mentioned in Chapter 4, under certain conditions instability implies ergodicity, which means that when unstable systems are concerned we should not worry too much about the missing information because ergodicity implies that any additional information will be, almost always, asymptotically irrelevant. Therefore, instability provides a nontrivial reason for accepting ergodicity. And such a reason may be formulated even within our doubly subjectivist view![9]

This argument, we believe, is an interesting one, because it proves that one may explain in subjectivist terms why, in some cases, the inductive method may fail. The reason is the epistemic characterization of Lagrange instability. That is, our measurements are not refined enough to allow us to determine, at present, all of the distinctions that may prove relevant in the future. Nevertheless, the argument we presented leads to a potential problem. We outlined a way of introducing a purely epistemic concept of instability, that is, instability that is not assumed to be generated by an intricate and sensitive underlying process; such a concept should be investigated further. On the one hand, an epistemic concept of instability may prove quite useful in areas where the very notion of an underlying process is hypothetical and abstract. (Such is the case in economics, for example.) On the other hand, it seems that in the context of SM epistemic instability is, in itself, a phenomenon to be explained. One feels that the doubly subjective view simply dodges the real problems. Therefore, we wish to suspend the verdict on the doubly subjectivist view until further studies are made.

---

[9] Notice that, within the extended subjectivist framework that we develop in this chapter, we may also clarify the difference between ultrasubjectivism and permissive subjectivism. The ultrasubjectivist does not have any use for the notion of stability, even when it is construed subjectively. The permissive subjectivist, on the other hand, may use the notion of instability to formulate various constraints on our degrees of belief. Such constraints can reflect the specific scientific background of various theories without smuggling in objectivist assumptions.

The second solution to the problems involved in the introduction of probabilities into the framework of classical mechanics is obtained by redefining the aims of the ergodic approach. Up until now, we have thought about the ergodic approach in terms of its emphasis on the centrality of the notion of time averages and the attempts made to connect them with phase averages. In the last chapter we contrasted this approach with the topological dynamic approach that investigates the asymptotic behavior of the individual trajectories. Now, we will try to combine these two ways of thinking into a single approach, which we will call *abstract dynamics*, and try to use it to construct new foundations to SM. The main tenet of the solutions offered by this approach is that many of the puzzles related to the notions of order and disorder, predictability and chance, periodicity and dispersion, and decomposability and homogeneity can be investigated and solved in a very abstract manner. Therefore, according to the present point of view, one should regard the discussion of these issues as part of rational mechanics, that is, that part of physics that is dedicated to the discussion of first principles. To be able to construct such an abstract solution, the first step is to "strip away" from classical mechanics all of the elements that are "inessential." That process ends with a presentation of physical systems that mentions only basic dynamic notions augmented by set-theoretic notions. We introduce a small number of probabilistic and topologic concepts within such a system. By using these constructs we can formulate very abstract versions of the ideas of Maxwell, Boltzmann, Poincaré, Birkhoff, and others. The claim is that, within this very elementary framework, we can arrive at satisfactory explanations of the emergence of equilibrium in particular and the phenomenon of deterministic randomness in general. This claim is supported at least partially by a group of powerful results that were obtained in the last three decades.

The dynamic approach was already being discussed in the middle of the nineteenth century. The development of these ideas was described by Maxwell, for example, in the Preface to his book, *Matter and Motion*.[10] "Physical science," he wrote, "which up to the end of the

---

[10] Originally published in 1877. In the subsequent footnotes, the page numbers will refer to the Dover edition.

eighteenth century had been fully occupied in forming a conception of natural phenomena as the result of forces acting between one body and another, has now fairly entered on the next stage of progress – that in which the energy of the material system is conceived as determined by the configuration and motion of that system, and in which the ideas of configuration, motion, and force are generalized to the utmost extent warranted by their physical definitions." The most elementary part of the dynamic approach is called *kinetics*, "the science of pure motion." "In kinetics," writes Maxwell, "the only attribute of matter which comes before us is its continuity of existence in space and time – the fact, namely, that every particle of matter, at any instance of time, is in one place and in one only, and that its change of place during an interval of time is accomplished by moving along a continuous path."[11] The basic assumptions of kinetics is that the systems involved are stationary and that space is homogeneous. "If causes differ only as regards the absolute time or the absolute space at which the event occurs, so likewise will the effects. (. . .) The difference between one event and another does not depend on the mere difference of the times or the places at which they occur, but only on the differences in the nature, configuration, or motion of the bodies concerned."[12] The next stages in the development of the science of motion are kinematics and dynamics. "When the mutual action between bodies is taken into account, the science of motion is called kinematics, and when special attention is paid to force as the cause of motion, it is called dynamics."[13] The aim of dynamics is to derive the laws governing the motion of a system of material points in terms of their configuration and their momentum. Given each of the possible initial conditions of the system, the laws of motion determine a trajectory in the phase space of the system. An abstract dynamic approach attempts to describe the system of possible trajectories in the most abstract way possible.

In Chapters 3 and 4 we introduced a very abstract dynamic description of a system of material points M. The description was given in terms of a structure $\langle S, B, W, T \rangle$.[14] As you may recall, S represents a set of possible states of M, B stands for the σ-algebra of measurable events, and T is a transformation that represents the law that governs the

---

[11] Matter and Motion, p. 78.
[12] Ibid, p. 13.
[13] Ibid, p. 26.
[14] For a survey article on related issues, see Nadkarni, "Descriptive Ergodic Theory," *Am. Math. Soc.* 94 (1989).

motion of M. The novel element in the definition was W, the σ-ideal of "small" sets. In what follows, we shall attempt to express some of the properties of M in terms of such abstract structures. Recall that abstract dynamic systems have both probabilistic and topological realizations. These are obtained by using the σ-ideal of null sets and the σ-ideal of sets of first category as W, respectively. Every property is investigated, as usual, modulo the members of W. This means that if a property holds for all but the members of a set in W, we regard the property as one that obtains almost always.

We assume, as we usually do, that T is a measurable transformation. In what follows, we shall also assume that T is W–quasi-invariant transformations, which means that $T(W) = W$. Quasi-invariance is an abstract variant of nondissipativeness or incompressibility. If A is a "large" set, it will not "become small." That is, if A is not in W, $T(A)$ will not be in W either.

In the present context we will restrict our attention only to a subset of the abstract systems, namely, to bounded systems. Such systems are characterized in the following way. As you may recall, A is a wandering set if $A \cap T^n(A) = \emptyset$ for every n. Therefore, if A is a wandering set of a "nonnegligible" size and T is W–quasi-invariant, then $\cup_{n=1}^{\infty} T^n(A)$ has to have an infinite size. Therefore, the set S that includes $\cup_{n=1}^{\infty} T^n(A)$ cannot be bounded. This insight is used to characterize boundedness. A system is bounded if and only if every wandering set is a member of W. (Notice that the definition of a σ-ideal implies that, when $A \in W$ and T is W–quasi-invariant, the set $\cup_{n=1}^{\infty} T^n(A)$ is a member of W as well.) Bounded systems generalize both the notion of a system on which a normalized probability may be defined and the notion of compactness. They are the abstract analogs of finite and closed systems. In particular, a system of gas molecules trapped in a container is a bounded system.

Bounded systems have a property that generalizes Poincaré's notion of recurrence. For every $A \in B$ there is a measurable $A_0 \subseteq A$ such that every $s \in A_0$ "returns" to A_0 infinitely often and $A \backslash A_0 \in W$. [$s \in A$ returns to A if, for some n, $T^n(s) \in A$.] The proof of this claim is simple. The set of elements that never return to their origins is a wandering set, and, hence, when bounded systems are concerned, it is a member of W. In fact, bounded systems have an even stronger property that is related to Khinchin's notion of recurrence. The set of "recurrence indexes" of the members of A_0 is relatively dense. (This means that the time intervals between consecutive returns are uniformly

202

bounded.) Moreover, we may also obtain a property that is related to Birkhoff's notion of recurrence. The relative frequency of recurrence in bounded systems is well-defined almost always. That is, $\lim_{n \to \infty} 1/n \Sigma_{n=1}^{\infty} X_A(T^n(s))$ is a well-defined quantity for almost every member of A.) The relative frequency of recurrence in A is, of course, proportional to the "size" of A.

Observe that the truth of these abstract theorems depends only on purely set-theoretic properties. So far, we have not even introduced probabilistic and topological concepts. Nevertheless, these theorems apply to any "real" system of material points. In particular, the universe as a whole, if it is taken to be a bounded system, will have recurring states. This result demonstrates the fact that powerful results with physical applications can be obtained in abstract dynamics.

To have a more general notion of relative frequency we have to redefine the notion of ergodicity. The new definition is the following. $\langle S, B, W, T \rangle$ is ergodic iff $T(A) = A$ implies that $A \in W$ or $S \backslash A \in W$. Ergodic systems have the property that, given any $A \in B$, the relative frequency $\lim_{n \to \infty} 1/n \Sigma_{n=1}^{\infty} X_A(T^n(s))$ is well defined for almost every s in S. Ergodicity functions in this abstract context as a weak characterization of systems that remain in an equilibrium state.[15] This means that if A is a "large set" (that is, if $S \backslash A \in W$,) then most trajectories will not fluctuate and stray too often from A. (If they strayed too often, their time averages would be skewed.)

If a system is not ergodic, it can be decomposed into its ergodic components and the decomposition may be affected in the following way. If S is not ergodic, then for some $A \in B$, $T(A) = A$, but neither A nor $S \backslash A$ are members of W. If A does not contain invariant subsets, we may regard it as an ergodic component. If it contains an invariant subset A', we should find out whether A' includes further invariant subsets. If it does not, we may regard A' as an ergodic component; otherwise, we should continue. The same procedure should be applied to $S \backslash A$ as well. When this process is terminated we will have an ergodic decomposition of S. The ergodic decomposition theorem allows us to think of each ergodic component as a separate ergodic system. Therefore, we may assume, without limiting the generality of the discussion, that we may

---

[15] As we remarked in Chapter 2, we need to introduce the notion of mixing to obtain a more satisfactory characterization of equilibrium. Unfortunately, there is no general decomposition theorem that guarantees that the different components are mixing.

concentrate only on ergodic systems and analyze nonergodic systems as if they were composed of a number of distinct ergodic subsystems. Therefore, we can state that the objects of our investigation are bounded quasi-invariant ergodic systems. Such systems constitute an exceedingly abstract representation of closed mechanical systems that do not have nontrivial invariants of motion.

The abstract version of the ergodic decomposition theorem is rather important conceptually. The theorem shows, without mentioning probabilistic or topological notions, that the very fact that we are dealing with bounded systems is sufficient to argue that, when all of the "invariants of motion" are taken into account, we obtain a description of the bounded system as an aggregate of ergodic systems that do not contain any further invariants of motion. This means that when we find out, in concrete cases, the identity of the ergodic component in which a system lies, there is nothing more to learn about the system. Each ergodic component is completely "homogeneous."

Another fact worth noting is that ergodicity is not only compatible with recurrence but, in fact, it implies it. This shows that our weakest notion of order, namely, recurrence or Poisson periodicity, is intimately related to the weakest notion of disorder, that is, to ergodicity. The argument is simple. As long as there is no constraint on the motion that will prohibit the system from visiting a given set ever, it will do so; therefore, if it visited the set once, there are obviously no such constraints, and, hence, it will visit the set again. Ergodicity, then, is in fact a rather weak notion.

The last points, it may be argued, have far-reaching consequences to our understanding of SM. One may argue that we have in fact demonstrated that SM should not be construed as a theory that deals primarily with many-particle systems.[16] It should be formulated as the theory that describes the behavior of any system that cannot be analyzed further.[17] The idea is that, when we approach a system, we first have to find out what are its invariants of motion, that is, which prop-

---

[16] We do not mean to imply that the number of degrees of freedom is unimportant. The number of degres of freedom enters the picture, for example, because the length of Poincaré's cycles is proportional to the number of degrees of freedom. (See Mazur and Montroll, "Poincaré's Cycles, Ergodicity, and Irreversibility in Assemblies of Coupled Harmonic Oscillators," *J. Math. Phys.* 1(1) (1960).

[17] Two important points should be noted. First, when systems of coupled harmonic oscillators are concerned, the distinction between decomposable and nondecomposable systems has a clear physical significance because decomposable systems

erties impose substantive restrictions on the set of states that are accessible to the system given its initial conditions. This investigation, which is the province of exact mechanics, will be terminated at a certain point. After that, we will begin a new kind of investigation, namely, we will concentrate on those properties that may be deduced from the assumption that all of the relevant invariants of the motion have been already identified and analyzed.[18] This type of investigation, it may be claimed, lies at the very foundation of SM.

This conception, as interesting as it may seem, raises some serious difficulties. The most urgent one, it seems, is the extent to which we trivialize SM when we formulate it in such an exceedingly abstract manner. It may be objected that the most interesting and informative properties of the systems that form the natural domain of SM are not determined by their abstract structure. Physical systems have many special properties; their laws of motion are represented by differential equations, and their evolution is described by diffeomorphisms of manifolds that preserve measures with smooth densities. Therefore, one may form the belief that SM proper begins only when we identify and characterize those features of "real" physical systems that distinguish them from other, more abstract systems. However, recent results prove that this is not the case. Let us then end this section with a brief discussion of some of these results.

Let us return to the structure $\langle S, B, W, T \rangle$. This time, though, we will introduce further assumptions. First, we will assume that among the members of B we may fix a countable set τ of "open neighbor-

---

will remain so even if they are slightly perturbed (a consequence of the KAM theory). Second, the distinction is not merely a conceptual one; we know, for example, how to transform a decomposable system into a nondecomposable one. When the energy level increases decomposable systems undergo a "stochastic transition" and become indecomposable. (In this context, this means that the fraction of the phase space occupied by invariant tori decreases.) For a discussion on the notion of stochastic transition, see O. Penrose, "Foundations of Statistical Mechanics," *Rep. Prog. Phys.* 42 (1979), pp. 1952–1955.

[18] We would like to remind the reader that the proof of the ergodic decomposition theorem depends, crucially, on our three assumptions, namely, that we have a fixed notion of "negligibility," that the negligible sets form a σ-ideal, and that the notion of negligibility is time-invariant. When these assumptions are questioned, and in many contexts they should be questioned, we may never be able to arrive at a stable decomposition. In such contexts it might be the case that (i) one may always be able to find new invariants of the motion, (ii) a combination of negligible factors might cease to be negligible, and (iii) factors that might seem negligible at one point in time will become important later.

hoods" that are sufficient to generate B and whose union covers S. We will also assume that T is continuous relative to τ; that is, if $A \in \tau$, then $T^{-1}(A) \in \tau$. (These assumptions guarantee that we may treat $\langle S,B,W,T \rangle$ as a standard Borel space.) The introduction of a topological structure marks a departure from the purely abstract framework that we have been preoccupied with thus far. It is remarkable, though, that this minimal topological structure allows us to characterize the members of W in probabilistic terms. Let $M_E\langle S,B,W,T \rangle$ be the set of ergodic quasi-invariant probability measures on $\langle S,B,W,T \rangle$. For every $m \in M_E\langle S,B,W,T \rangle$, let N_m be the set of m-null sets.

Theorem I (Weiss and Shelah, 1982).[19] $W = \cap_{m \in M_E} N_m$.

In other words, the members of W are exactly those "universally null" sets. Observe, however, that different ergodic measures are mutually singular to one another; hence, it must be the case that $M_E\langle S,B,W,T \rangle = \{m\}$. Systems that have only one ergodic measure are called *uniquely ergodic systems*. The following theorem shows that uniquely ergodic systems are not the exception, but, in some sense, the rule.

Theorem II (Jewlett, 1970; Krieger et al., 1972[20]). *Every ergodic measure preserving system on a Lebesgue space is measure-theoretically isomorphic to a minimal, uniquely ergodic homeomorphism of a compact metric space.*

As we remarked earlier, whenever we can demonstrate that a system is both ergodic and minimal, we obtain objects with very special properties. On the one hand, minimality guarantees that *every* trajectory approaches arbitrarily close to every state. On the other hand, ergodicity implies the fact that the time averages taken relative to the different trajectories are almost always the same. Therefore, in such cases the idea of thinking of the probabilities as time averages has a dual justification, corresponding to both the RT assumption and the hypothesis of MT. This dual feature guarantees a harmony between the two types of inductive inferences. We may study such systems by using the methods of ergodic theory and concentrate on their average dynamics, or we may investigate the asymptotic properties of their individual tra-

---

[19] See Weiss and Shelah, "Measurable Recurrence and Quasi Invariant Measures," *Israel. J. Math.* 43 (2) (1982).
[20] For a discussion on this theorem, see Peterson, *Ergodic Theory*, Cambridge Univ. Press (Cambridge, 1994) section 4.4, p. 186.

jectories by using the methods of topological dynamics; both lines of inquiry complement each other. Therefore, the fact that, in some theoretical sense,[21] such objects are not "exceedingly rare" has an important philosophical consequence: Our two versions of inductive reasoning are harmonious with one another. Petersen also conjectures[22] that every ergodic measure-preserving transformation is isomorphic to a diffeomorphism that preserves a smooth measure on a manifold (subject, perhaps, to the condition of finite entropy). If this were the case, then the result is that much of what is unique to physical systems is contained, implicitly, in their abstract topological structure![23]

Another different, but related set of remarkable results was obtained by D. Ruelle. The general aim of Ruelle's investigation was, again, to investigate physical systems in the most abstract way possible. Ruelle used a very abstract notion of topological entropy and managed, by utilizing a variational principle, to show that much of Gibbs's theory may be derived in an extremely abstract manner.[24]

5.4. PRAGMATIST FOUNDATIONS OF SM

The third solution we will discuss will be called the pragmatist solution, although the term solution might be misleading. The pragmatists are not very troubled by the fact that probabilities cannot be defined from other physical parameters. They simply do not believe that it is necessary to give every concept that appears in a physical theory a physicalist interpretation. This view is often motivated by a broader outlook. Pragmatists believe that we have a total freedom to choose the concepts that appear in our scientific theories. We do not have to

---

[21] Petersen writes that "the Jewlett Krieger theorem is much like the isomorphism theorem for Lebesgue spaces: its theoretical importance is tremendous, but its practical effects are practically nil. Indeed, the theorem, itself, says that this must be so, since it guarantees that the property of being realizable as a uniquely ergodic homeomorphism tells us nothing at all about a given measure-preserving transformation" (ibid, p. 187).

[22] See ibid.

[23] This point, however, should not be overstated. Hamiltonian systems are mathematically very unique, and they contain structural features that cannot be encoded by their abstract structure. For example, the "smoothness" of the physical motion is extremely important. Another example is the symplectic structure of the phase space. Nevertheless, mathematicians were genuinely puzzled by the theorem that we cited.

[24] See Ruelle, *Thermodynamical Formalism*, Addison-Wesley (Reading, MA, 1978).

justify the initial choice of concepts at all.[25] All we need is to make sure that the theories we construct will imply correct predictions. Once a theory is proven correct, we may treat it as if it were true. This means that we may assign to its concepts standard semantic interpretations. In other words, we may treat the concepts as if they were physically real. An example of a pragmatist's treatment of a concept is W. James's conception of God. James argued that rather than deciding, at the outset, whether the concept of God is a referring term or not, one should investigate whether including God in one's world picture is beneficial to one's life as a whole. If it is, one may behave as if God were a referring term. This basic idea, we believe, is analogous to Gibbs's conception of SM.

We have already had an opportunity to review Gibbs's SM in Chapter 1, where we tried to explain why Gibbs's theory, as much as it was confirmed empirically, was considered at the time to be somewhat unsatisfactory. Since the publication of *Elementary Principles*, unceasing efforts were invested in attempts to "correct" Gibbs's theory. We have seen how Jaynes, for example, building on earlier work by Tolman and ter Haar, introduced the canonical distribution by a variational principle that was elevated to the status of a principle of reasoning. This idea, though, is not a correct interpretation of Gibbs's own views.[26] Gibbs's approach to SM was a pragmatist one, and we believe that his methods and reasoning were thoroughly different from Jaynes, that is, insofar as Jaynes was a subjectivist. We believe, however, that Jaynes, too, had pragmatist intuitions and that sometimes he was more like a pragmatist than like a subjectivist. Let us return to the discussion of Gibbs's writing and attempt to understand his views a little better.

Gibbs's point of departure was that SM is the theory of systems with many degrees of freedom. Such systems differ from simpler mechanical systems *only* in their number of degrees of freedom. The laws that govern such systems are the same mechanical laws that govern the motions of the planets, for example. Nevertheless, Gibbs argued, there is no reason why we should not present the theory of systems with many degrees of freedom ex novo. Such a presentation,

---

[25] Compare with Rorty, *Consequences of Pragmatism*.
[26] Gibbs, of course, was the one who discovered the variational principle behind MEP (see ibid, ch. XI). It seems, though, that for Gibbs the principle was a consequence of the definition of his canonical ensembles rather than the definition itself.

according to Gibbs, is simply more convenient. This point is simple enough, but behind it lies a revolutionary methodology: A physicist may introduce a theory for the sake of convenience without making any fundamental claims about the unity or the ontological distinctness of the underlying domain of inquiry. This point is what we call Gibbs's pragmatism.

Taken as the theory of systems with many degrees of freedom, SM clearly has one leg in mechanics and the other in statistics. Gibbs's main discovery was his way of combining the two. The combination was affected in the following way. In the first chapter, Gibbs introduced a mechanical principle, "the principle of the conservation of probability in time," which applies to all mechanical systems, independent of their number of degrees of freedom. (This principle is, in fact, a consequence of Liouville's theorem.) In the second chapter, he introduced a statistical principle, the theory of errors, which applies to any system (not necessarily a physical system) that involves a large number of parameters. These two principles, taken together, were shown by Gibbs to be sufficient for the introduction of probabilities into SM.

Gibbs reasoned in the following manner. Suppose that we are trying to evaluate the probability $P(s_t \in A)$. Because M is a mechanical system, we assume that $P(s_t \in A)$ is dependent only on the dynamic variables of M. Therefore, to calculate $P(s_t \in A)$ for an arbitrary A, it is sufficient to solve $P(s_t \in E_i)$, where E_i are intervals of position and momentum. On the other hand, Gibbs tells us, "we suppose, as is usual in the 'theory of errors' that the determination (explicit or implicit) of the constants of the motion is of such precision that the coefficient of the probability e^η or $e^{-F} \ldots$ is practically zero except for very small values of the differences $p'_1 - P'_1, q'_1 - Q_1, \ldots$ etc."[27] ($P'_1, Q'_1, \ldots$, etc., were defined as the most probable values and $p'_1, q'_1, \ldots$, etc., are the "fluctuations.") Next Gibbs looked again at the system as a mechanical system. "We shall suppose," writes Gibbs, "that the forces to which the system is subject are functions of the coordinates alone or with time. The principle of the conservation of the probability of phase will therefore apply, which requires that at any other time t'' the maximum value of the coefficient of probability shall be the same as at time t', and that the phase $(P'', Q'', \ldots$ etc.) which has the greatest probability will correspond to the phase $(P'_1,$

[27] Gibbs, *Elementary Principles*, p. 21.

$Q'_1, \ldots$ etc.)."[28] More generally, he writes, "the probability that the phase at t' falls within the limits $F'' = k$ is the same as the probability that it falls within the limits $F'' = k$ at the time t'', since either event necessitates the other."[29]

Note that this property is always satisfied when we assume that the behavior of the system at t' is independent from the behavior at t''. For example, if we make two independent series of measurements on the same system, we expect the error distribution to be the same in both cases. It took Gibbs's ingenuity, though, to conjecture and prove the fact that the same can be true even for systems whose evolution is completely deterministic!

The above-mentioned considerations constitute the necessary background for the introduction of the canonical distribution. Gibbs characterizes it as the "most simple case" that has the following property:

$$\sum_i \left(\frac{DP}{dp_i} p_1 + \frac{DP}{dq_1} q_1 \right) = 0. \tag{*}$$

This property, which expresses the fact that the distribution will not change in time, was taken by Gibbs to be the definition of statistical equilibrium. A necessary and sufficient condition for the satisfaction of (*) is that "P (the probability) should be a function of the p's and q's," (the momenta and coordinates) "which do not vary in time in a moving system."[30] Gibbs demonstrates the fact that the canonical distribution has property (*) by the following argument:

The distribution represented by . . . $P = e^{\psi - \epsilon/\theta}$ (e is the energy, f and q are constants) seems to represent the most simple case conceivable since it has the property that when the system consists of parts with separate energies, the laws of the distribution in the phase of the separate parts are of the same nature, a property which enormously simplifies the discussion and is the foundation of extremely important relations to thermodynamics.[31]

[This property, of course, implies (*). The canonical distribution, when restricted to any subensemble, remains essentially the same. This fact, in and of itself, implies that we cannot improve our predictions by taking into account additional information.] Concerning the interpretation of the canonical distribution, Gibbs makes the following remark:

[28] Gibbs, *Elementary Principles*, p. 21.
[29] Ibid, p. 23.   [30] Ibid, p. 32.   [31] Ibid, p. 33.

The fractional part of an ensemble canonically distributed which lies within any given limits of phase is therefore represented by the multiple integral

$$\int\!\!\int \ldots \int e^{\psi-\varepsilon/\theta} dp_1 \ldots dq_n$$

taken within those limits. We may express the same thing by saying that the multiple integral expresses the probability that an unspecified system of the ensemble (i.e., one of which we only know that it belongs to the ensemble) falls within the given limits.[32]

Notice that Gibbs's justification of the choice of the canonical distribution has a thoroughly pragmatic character. The canonical distribution is the "most simple case conceivable" and it "simplifies the discussion enormously." This characterization has mystified many generations of physicists ever since. Many readers of Gibbs asked themselves, How could SM probabilities, which, as Gibbs insisted, are determined by the dynamic variables, be viewed pragmatically?

It is interesting to note that these problems with which physicists occupied themselves in the beginning of the twentieth century are essentially problems that were raised and solved a century earlier by mathematical statisticians.[33] These problems concern the status and justification of the theory of errors in general and the normal curve in particular. The solution that was obtained by Laplace on the basis of the work of Gauss was to justify the theory of errors using a new theorem, namely, the central limit theorem. The solution was known as early as 1810. Nevertheless, it was only in the 1940s that Khinchin understood that the central limit theorem lies at the very basis of SM as well. In other words, it took the community of physicists almost 130 years to learn how to make use of the ingenious and fairly well known work of the fathers of mathematical statistics!

The problem of the justification of the assumption that the normal distribution represents the error curve occupied many statisticians in the beginning of the nineteenth century. We have already seen in Chapter 1 how Herschel, who tried to explain the ideas of Quetelet, attempted to derive the normal curve from a series of simple assumptions, and how Maxwell applied the same procedure to the field of statistical physics. We also learned that Herschel's assumptions, as natural as they sound, are too restricting. For example, the idea

---

[32] Ibid, p. 34.
[33] See Stigler, *The History of Statistics*, Belknap (Mass, 1986), for a history of the subject.

that each measurement should be presented as the composition of independent components led to the ultimate failure of Maxwell's first attempt to justify his ideas about the distribution of velocities. These problems, however, were solved by Laplace in 1810, and the solution was certainly known to Quetelet in the 1840s.[34] The basic idea behind the solution was contained in the central limit theorem. Stated plainly, the content of the theorem is the following. Suppose that we are dealing with a large number of measurements and the results of each measurement are determined by a large number of independent factors whose effects are relatively small. In such a case, the resulting distribution of errors may be approximated by the normal curve independentally of the precise character of the determining factors. The central limit theorem, unlike earlier attempts to justify the theory of errors, depended only on a single "a prioristic" assumption, namely, one needs to suppose that the error curve is symmetric. Therefore, the theorem seems almost miraculously strong. As Stiegler puts it, the theorem generated a feeling of wonder. Out of chaos came a sense of order. "A simple component of the measurement, call it error if you like, may be totally unpredictable; but the average of many such components, no one or few dominant (that is, all subject to the same general dominating forces), will be subject to a law of a remarkably specific form."[35] Indeed, the central limit theorem is generally considered to be one of the most important mathematical results obtained in the nineteenth century.

The central limit theorem should have attracted the attention of mathematical physicists. It seems, at least prima facie, to be ideally suited for applications to SM, a theory that, by Gibbs's own definition, concentrates on systems composed of many particles. The idea should have been the following. Suppose that we make a large number of measurements on the various members of an ensemble. Each of these measurements is determined by a large number of factors, the contribution of which is almost negligible (these factors are, of course, the individual molecules). Therefore, an application of the central limit theorem could have been attempted, that is, if we could think of the effects of the individual molecules as being independent events. But

---

[34] See Stigler, Chapters 3 and 4, for a description of the development of the ideas leading to the central limit theorem and its relation to the justification of the theory of errors.

[35] Ibid, p. 203.

even if the idea occured to someone, he or she would have had to over-come a substantive obstacle to apply the theorem. The individual mol-ecules do not act independently. On the contrary, the system as a whole is governed by completely deterministic laws. This difficulty, then, pre-vents a straightforward application of the central limit theorem to sta-tistical mechanics.

The solution to the problem was obtained by Khinchin. He observed that, even if the various particles are not independent, they are never-theless "approximately independent" in the sense of having an expo-nential decay of autocorrelation: Even an exact knowledge of the state of an individual particle quickly becomes almost completely useless for predicting the future states of this particle. This weaker property, then, was shown by Khinchin to be sufficient for the application of the central limit theorem. More precisely, Khinchin argued in the follow-ing way:

Because of the fact that the given system consists of a very large number of molecules it is natural to expect that the knowledge of the state of a single molecule at a certain moment does not permit us to predict anything (or almost anything) about the state in which this molecule will be found after a sufficiently long time. For example, the exact knowledge of the energy of a given molecule at a given moment of time cannot give us any indication of the value which this energy will have several hours later (because of the large number of collisions suffered by the molecule during this time interval). This statement seems to us so natural that it would be difficult to think otherwise; in fact, this represents the basic idea of "molecular chaos."[36]

Notice that this assumption may not be true for all of the particles, and hence its exact formulation necessitates the introduction of a statisti-cal element into the argument. But once the assumption was made, Khinchin was free to apply the central limit theorem. By using the theorem, Khinchin derived the microcanonical distribution and from it, by using another approximation that depends only on the same set of assumptions, he obtained the canonical distribution.[37]

---

[36] See Khinchin, *The Mathematical Foundations of Statistical Mechanics*, Dover (London, 1949), p. 67.

[37] See the discussion in Chapter 2 for more details. Khinchin's results depended on the assumption that, in the context of SM, all of the functions we are dealing with are "sum functions." These results were generalized by Albertoni, Bochieri, and Loringer, "New Theorem in Classical Ensemble Theory," *J. Math. Phys.* 1, 244. (1960). Similar results were proven for quantum SM by Prosperi and Scotti, "Ergodicity Conditions in Quantum Mechanics," *J. Math. Phys.* 1, 218 (1960).

Finally, we are in a position to present the Gibbs–Khinchin formulation of SM as a pragmatist approach. The basic idea is simple. SM is defined as the study of mechanical systems with many particles whose individual effects on the system as a whole are negligible.[38] It is of course clear that statistical mechanics overlaps with both mechanics and thermodynamics. Moreover, many-particle systems do not constitute a distinct new "physical type" of system. Therefore, it is somewhat artificial to try to deduce special laws for such systems. (It may also be maintained that we need to prove that the Khinchian version of the assumption of molecular chaos is consistent with the underlying dynamics!) Consequently, physicists with a metaphysical bent might object to the idea of introducing SM as a new physical theory. According to the pragmatist, though, such worries should not trouble the physicist provided that the resulting theory is simple and useful. The only criterion one should observe is the pragmatic principle, namely, one should ask whether it is a fruitful idea to treat the collection of many-particle systems as constituting a coherent object of study.

Once we delineate the subject matter of SM in this way, we are guaranteed to be able to apply the central limit theorem and, hence, the theory of errors. Consequently, we may use the Maxwell–Boltzmann distribution to calculate the probabilities of various events (or, alternatively, we may base our predictions on the canonical distribution). Notice that the assumption of the exponential decay of correlation implies ergodicity,[39] so we do not have to worry about the effects of new information on our calculations.

5.4. A CONCLUSION

Rationalists believe that the book of nature is written in the language of mathematics. This belief has far-reaching consequences for their conception of physics: Insofar as physical theories are formulated in mathematical terms, they have a capacity, by virtue of their mathematical structure, to "mirror" the natural processes that they describe in a transparent manner.

---

[38] A consequence of this assumption is that the particles are uncorrelated. Otherwise, their combined effect, and hence their individual effects, could be large. A property that expresses the fact that the various particles are not correlated is called by Jancel "weak coupling." Jancel, ibid, p. 332.

[39] See Chapter 2.

214

The rationalist's conception of physics seems quite justified in certain contexts. One would be hard pressed to point out aspects of the physical world that the second law of Newton, for example, leaves out or aspects of the formulation of the second law that are superfluous or arbitrary. More generally, the language of the theory of differential equations seems ideally suited for describing how the various physical parameters determine the evolution of physical systems. Everything that matters is expressed; nothing in the form of expression is arbitrary or superfluous.

This, however, is not the case with SM. To be sure, everything that matters is taken into account. Otherwise, SM would not have been a correct physical theory. Therefore, although many details are left out of the discussion, we discard them because they do not really matter. This feature of statistical mechanics is rather extraordinary. It took the genius of Maxwell, Boltzmann, and Gibbs to manage to sieve through an immense number of facts and leave in their theories only the bare minimum, that is, only those facts that truly matter. The problem (if it is indeed one) is that the statistical concepts seem to be rather removed from or indifferent to the particular physical facts. They do not seem to be derived from the phenomena themselves, and hence they seem to "force themselves" on the phenomena. This sentiment is familiar in the context of the discussion on the application of statistics to the social sciences. The statistical averages seem to be indifferent to the life of particular individuals, to the fact that they possess a free will, and to their relative autonomy. But, on reflection, it should be obvious that, even if particles do not possess a free will, they do have a measure of relative autonomy and they do create a specific constellation of facts. And the statistical laws are somehow foreign to these facts. Therefore, SM presents a difficulty to the rationalists. Even if SM is formulated in mathematical terms, it is not written in the language of nature; it is written in the language of human observers, reflecting their lack of precision, their limited computational capacity, and the methods that they use to overcome the effects of observational errors.

It is our view, then, that many aspects of SM seem puzzling to us because most of us have some rationalist predelections, and, because we do, we have certain expectations from physics. When these expectations are not met, we are baffled. Fortunately, most of us have some other sentiments as well. Therefore, when we are confronted with the fact that physics has a variety of theories of different types (all of which are correct and useful), we can come to terms with this fact without much difficulty.

The conceptions we outlined in this chapter are three different ways of coming to terms with the "human" side of SM. The subjectivist's conception takes the philosophical challenge head on. War is declared on rationalism, on determinism, and on the very idea that we can "read the book of nature." The abstract dynamic approach advocates the more traditional response to the rationalists, namely, it points out the fact that, in general, even the most objective theories require a construction of a conceptual "scaffolding" on which our empirical theories can be built. Finally, the pragmatist, for whom traces of human will and interest are to be found everywhere, tells us precisely how to transform the language of nature, that is, how to introduce a new category of systems that will enable us to regard SM as the theory that describes the laws that govern systems belonging to this new "unnatural" category.

These three conceptions are all different, and there is no reason to believe that we have exhausted all of the possibilities. What is common to all of them is a rejection of extreme rationalism. This rejection should be welcome. It seems to us to be a prerequisite for a more reasonable and less dogmatic view of science.

Appendix I

Mathematical Preliminaries[1]

I.1. ELEMENTARY MEASURE THEORY

Let S be a set. A collection B_0 of subsets of S is called an *algebra* iff:

(i) If $A, A' \in B_0$, then $A \cap A' \in B_0$ and $A \cup A' \in B_0$.
(ii) If $A \in B_0$, then $S \backslash A \in B_0$.
(iii) $S \in B_0$.

A collection B is called a *σ-algebra* if it is an algebra and, in addition,

(iv) if $\{A_n\}$ is a countable collection of subsets of S such that, for every n, $A_n \in B$, then $\cup_n A_n \in B$.

A *measurable space* is a structure $\langle S,B \rangle$, where S is a set and B is a σ-algebra of subsets of S.

Let $\langle S,B \rangle$ be a measurable space. A *finitely additive measure* is a function $m : B \rightarrow [0,\infty)$ such that:

(i) If $A,A' \in B$ and $A \cap A' = \varnothing$, then $m(A \cup A') = m(A) + m(A')$.
(ii) $m(\varnothing) = 0$.

If $m(S) < \infty$, m is a *finite measure*. If $m(S) = 1$ m is a normalized probability measure.

m is a measure if it is a finitely additive measure and, in addition,

(iii) If $\{A_n\} \subseteq B$ is a countable collection of mutually exclusive measurable sets, then $m(\cup_n A_n) = \Sigma_n m(A_n)$ (σ-additivity).

$\langle S,B,m \rangle$ is called a *measure space*.

---

[1] For further information, proofs, or references to proofs, the reader may consult Walters, *An Introduction to Ergodic Theory*, Springer–Verlag (New York, 1982).

Theorem (Caratheodory). *If* m *is a σ-additive measure on an algebra* B_0, *it can be extended, uniquely, to a measure on the σ-algebra that extends* B_0.

A collection of sets C is a *σ-ideal* iff:

(i) If $\{A_n\}_{n=1}^{\infty} \subseteq C$ is a countable collection of measurable sets, then $\cup_{n=1}^{\infty} A_n \in C$.
(ii) If $A \in C$ and A' is any measurable subset, then $A \cap A' \in C$.

Theorem. *Let* $\langle S,B,m \rangle$ *be a measure space. The collection of measurable sets with* m-*measure zero is a σ-ideal.*

I.2. MEASURABLE FUNCTIONS AND INTEGRATION

Let $\langle S,B,m \rangle$ be a measure space and $A \in B$. The *indicator function* of A is defined as $X_A : S \to \{0,1\}$ such that $X_A(s) = 1$ if $s \in A$ and $X_A(s) = 0$ otherwise.

We shall define the *integral* of X_A to be $\int X_A\, dm = m(A)$.

By an obvious generalization we shall define $\int f dm$, where f is a function that takes a finite set of values $r_1, \ldots, r_n$ as $\int f dm = \sum_{n=1}^{\infty} r_n m$ ($\{s \in S \mid f(s) = r_n\}$).

Theorem. *Let* f_n *be a sequence of functions such that, for each* n, $\int f_n\, dm$ *is well defined. If* $f_n \to f$ *uniformly, then* $\int f\, dm$ *exist and* $\int f_n dm \to \int f\, dm$.

I.3. TOPOLOGICAL SPACES

A *topological space* $\langle S, A \rangle$ is a structure where S is a set and A is a set of subsets of S that are called "open sets." We assume that A has the following properties:

(i) $\varnothing \in A, S \in A$.
(ii) If $\{A_\alpha\} \subseteq A$, then $\cup A_\alpha \in A$.
(iii) If $A_1, \ldots, A_n \in A$, then $\cap A_n \in A$.

$\langle S, A \rangle$ is called a *Hausdorffspace* if, for every $s, s' \in S, s \neq s'$, there are $A_s, A_{s'} \in A$ such that:

(i) $s \in A_s$ and $s' \in A_{s'}$.
(ii) $A_s \cap A_{s'} = \varnothing$.

218

The definition of Hausdorff spaces uses a *separation axiom*. A weaker separation axiom is the following. For every $s, s' \in \mathbf{A}$ there is $C \in \mathbf{A}$ such that $s \in C$ and $s' \notin C$.

An open cover for S is a collection $\{A_\alpha\} \subseteq \mathbf{A}$ of open sets such that $\cup A_\alpha \supseteq S$.

$\langle S, \mathbf{A} \rangle$ is a metrizable space if it is a topological space and if every open cover $\{A_\alpha\}$ has a countable subcover $\{A_n\}_{n=1}^\infty \subseteq \{A_\alpha\}$.

A *metric* $d : S \times S \to [0, \infty)$ has the following properties:

(i) $d(s, s) = 0$.
(ii) $d(s, s') = d(s', s)$ for every $s, s' \in S$.
(iii) $d(s, s'') \leq d(s, s') + d(s', s'')$ (the triangle inequality).

Let d be a metric on S. Let $s_0 \in S$ and $\varepsilon > 0$. A subset $A(s_0, \varepsilon) = A_{s_0,\varepsilon}\{s \in S | d(s, s_0) < \varepsilon\}$ is called an *open neighborhood* of s_0.

Theorem. *Let $\langle S, \mathbf{A} \rangle$ be a metrizable topological space.*

(i) *It is possible to find a metric* d *on* S.
(ii) **A** *can be presented as a collection of unions and finite intersections of open neighborhoods relative to* d.

Let $C \in \mathbf{A}$. The set $S \backslash C$ is called a *closed* set.

Let $\{s_n\}$ be a countable sequence of elements of S and let $s \in S$. The sequence s_n converges to s iff $d(s_n, s) \to 0$ as $n \to \infty$.

A sequence $\{s_n\}_{n=1}^\infty$ is called a *Cauchy sequence* iff, for every $\varepsilon > 0$, there is an $N > 0$ such that, for every $n, m > N$, $d(s_n, s_m) < \varepsilon$. s is called the limit of s_n.

Theorem. *If* C *is a closed set and* $\{s_n\} \subseteq$ C *is a Cauchy sequence, then there is* s $\in$ S *such that* s_n *converges to* s.

Let $C \subseteq S$. The smallest closed set containing C is called the *closure of C* and is denoted by $\bar{C}$. The largest open set contained in A is called *the interior of A*, and we denote it by $\mathrm{int}(A)$.

A set C is dense in S iff $\bar{C} = S$.

Theorem. *If* A *is dense in* S, *then for every* s $\in$ S *and every* $\varepsilon > 0$ *there is* s' $\in$ A *such that* d(s,s') $< \varepsilon$. *(Dense sets "approximate" all of the elements of* S.)

A set C is called a *nowhere dense set* iff $\bar{C}$ does not contain any open neighborhoods.

Theorem. *The following are equivalent:*

(i) C *is nowhere dense.*
(ii) $\bar{C}$ *has an empty interior.*
(iii) $S \backslash \bar{C}$ *is a dense set.*

$\langle S, \mathbf{A} \rangle$ is called a *separable space* iff it has a countable dense set.

A set A is called a *set of first category* iff it can be expressed as a countable union of nowhere dense sets. If A is not of the first category, it is called *a set of the second category*.

Theorem. *If* A *is a set of the second category, it can be expressed as a countable intersection of dense sets.*

Theorem. *Sets of the first category form a σ-ideal.*

$\langle S, \mathbf{A} \rangle$ is a *complete space* if it contains the limits of all of its Cauchy sequences.

$d(A) = \sup_{s,s' \in A} \{d(s,s')\}$ is called *the diameter* of A.

S is called *a bounded space* iff $d(S) < \infty$.

$\langle S, \mathbf{A} \rangle$ is compact iff every open cover has a finite subcover.

Theorem. $\langle S, \mathbf{A} \rangle$ *is compact iff every infinite sequence has a converging subsequence.*

Theorem. *If* $\langle S, \mathbf{A} \rangle$ *is a compact metric space, it is complete and bounded.*

Theorem (Baire's category theorem). *If* $\langle S, \mathbf{A} \rangle$ *is a complete metric space,* S *is a set of the second category.*

Theorem (Lebesgue's covering theorem). *If* $\langle S, \mathbf{A} \rangle$ *is a compact metric space, then, for every* $\epsilon > 0$, *there is a finite open cover* $\{A_i\}$ $0 \le i \le n$ *such that every set* A *with* $d(A) < \epsilon$ *is contained in* A_i *for some* $0 \le i \le n$.

220

I.4. CONTINUOUS FUNCTIONS

Let $\langle S, \mathbf{A} \rangle$ and $\langle S', \mathbf{A}' \rangle$ be two topological spaces. A function $f: S \to S'$ is called *continuous* iff, for every open set $C \in A', f^{-1}(C) \in A$.

A continuous function f is called a *homeomorphism* iff it is 1–1 and its inverse is continuous.

Theorem. *If* $s_n \to s$ *and* f *is continuous, then* $f(s_n) \to f(s)$.

f is called *uniformly continuous* iff, for every $\in > 0$, there is $\delta > 0$ such that, whenever $d(s_1, s_2) < \in$, then $d'(f(s_1), f(s_2)) < \delta$.

f *is bounded* iff there is $N > 0$ such that, for every $s \in S, f(s) \leq N$. f *attains its bounds* iff, for some $s \in S, f(s) = N$.

Theorem. *Let* $\langle S, A \rangle$ *be a compact space, and let* f: S $\to$ S *be a continuous function.* f *is bounded and it attains its bounds.*

I.5. MEASURES ON COMPACT METRIC SPACES

Let $\mathbf{S} = \langle S, \mathbf{A} \rangle$ be a topological space. $\langle S, \mathbf{B} \rangle$ is called the *Borel measurable space* of $\mathbf{S}$ iff B is the σ-algebra generated by the open sets of A. A measure m defined on B is called a *Borel measure*.

$\langle S, \mathbf{B}, m \rangle$ is a *regular Borel space* iff $\langle S, \mathbf{B} \rangle$ is a Borel space and for every $C \in B$ $m(C) = \inf\{m(D) | D \supseteq C$ and D is open$\} = \sup\{m(D) | C \supseteq D$ and D is closed.$\}$ In such a case, m is called a *regular Borel measure*.

Theorem. *Let* $\langle S, A \rangle$ *be a metrizable space. If m is a Borel measure on the corresponding Borel space, m is a regular Borel measure.*

Theorem. *Let* $\langle S, A \rangle$ *be a compact metric space, and let* $\langle S, B \rangle$ *be the corresponding Borel measurable space. If m is a finitely additive measure on* $\langle S, B \rangle$*, it is also* σ-additive.*

Proof. Let $\{A_n\}$ $0 \leq n \leq \infty$ *be a countable collection of mutually exclusive sets. We shall prove that m* $(\cup_{n=1}^{\infty} A_n) = \Sigma_{n=1}^{\infty} m(A_n)$. *For each* A_n, *let* $C_n \supseteq A_n$ *be an open set such that* $m(A_n) = m(C_n)$. *(We may assume that such a set exists because m is a regular Borel measure and* $\langle S, B \rangle$ *is compact.) Because* $A_i \cap A_j = \varnothing$, $m(C_i \cap C_j) = 0$. *[Otherwise, let us choose* $D \supseteq (C_i \cap C_j)$ *such that* $m(D) = m(C_i \cap C_j)$ *and* D *is open. Clearly,* $C_i \cap D \supseteq A_i$, $C_i \cap D$ *is open and* $m(C_i \cap D) < m$ (C_i) *in*

221

contradiction to the fact that $m(C_i) = \inf\{C \in B \mid C \supseteq A$ and C is open\}.]

Let $A^* = \bigcup_{n=1}^{\infty} A_n$. Because $C_n \supseteq A_n$ for every n, $\bigcup_{n=1}^{\infty} C_n$ is an open cover of A^*. Therefore, for some finitely many $C_{n_1}, C_{n_2}, \ldots, C_{nm}$ the set $C_{n_1} \cup C_{n_2} \cup \ldots \cup C_{nm}$ covers A^*. (Again, this follows from compactness.) In other words,

$$\bigcup_{n=1}^{\infty} A_n \subseteq C_{n_1} \cup C_{n_2} \cup \ldots \cup C_{nm},$$

and hence,

$$m\left(\bigcup_{n=1}^{\infty} A_n\right) \leq m(C_{n_1} \cup C_{n_2} \cup \ldots \cup C_{nm}),$$

but m is finitely additive and $m(C_i \cap C_j) = 0$. Hence,

$$m(C_{n_1} \cup C_{n_2} \cup \ldots \cup C_{nm}) = m(C_{n_1}) + m(C_{n_2}) + \cdots + m(C_{nm}).$$

and

$$m\left(\bigcup_{n=1}^{\infty} A_n\right) \leq m(C_{n_1}) + m(C_{n_2}) + \cdots + m(C_{nm}).$$

But because, for every n, $m(A_n) = m(C_n)$, it is also the case that

$$m\left(\bigcup_{n=1}^{\infty} A_n\right) \leq m(A_{n_1}) + m(A_{n_2}) + \cdots + m(A_{nm}).$$

However, clearly,

$$m\left(\bigcup_{n=1}^{\infty} A_n\right) \geq m(A_{n_1} \cup A_{n_2} \cup \ldots \cup A_{nm}) = m(A_{n_1}) + m(A_{n_2})$$
$$+ \cdots + m(A_{nm}).$$

(m is finitely additive and $A_i \cap A_j = \varnothing$). Therefore,

$$m\left(\bigcup_{n=1}^{\infty} A_n\right) = m(A_{n_1}) + m(A_{n_1}) + \cdots + m(A_{nm}).$$

and hence

$$m(A^* \setminus \{A_{n_1} \cup A_{n_2} \cup \ldots \cup A_{nm}\}) = 0.$$

In other words, for every A_1 such that $A_1 \neq A_{n_i}$ for every n_i, we have

$$m(A_1) = 0.$$

The reason is, of course, that $A^* \setminus \{A_{n_1} \cup A_{n_2} \cup \ldots \cup A_{nm}\} \supseteq A_1$, and hence it is also the case that $m(A^* \setminus \{A_{n_1} \cup A_{n_2} \cup \ldots \cup A_{nm}\} \geq m(A_1))$. Hence,

$$\sum_{n=1}^{\infty} m(A_n) = 0 + 0 + \cdots + m(A_{n1}) + 0 + \cdots m(A_{n2})$$
$$+ 0 + \cdots + \cdots + (A_{nm}) + 0 + \cdots = m(\bigcup_{n1}^{\infty} A_n.)$$

I.6. TOPOLOGICAL GROUPS

A *group* is a structure $\langle S, o \rangle$ such that S is a set and $o : S \times s \to S$, the group composition, is a binary operation with the following properties:

 (i) o is 1–1. (The composition of elements is uniquely defined.)
 (ii) For some $s_0 \in S$, $s_0 \cdot s = s \cdot s_0 = s$. (There is a "zero" element.)
(iii) For every $s \in S$ there is $s' \in S$ such that $s \cdot s' = s_0$. (There are "inverses" defined for each element.)

Let $s, s' \in S$. If, in addition to (i)–(iii),

(iv) $s \cdot s' = s' \cdot s$

for every $s, s' \in S$, $\langle S, o \rangle$ is called a *commutative* or *Abelian group*.

Let $\langle S, A \rangle$ be a topological space. Let $o : S \times S \to S$ be a well-defined operation such that $\langle S, o \rangle$ is a group. $\langle S, A, o \rangle$ is called a *topological group* iff o is continuous. (An open set in the space $\langle S \times S, A \times A \rangle$ is a set of the form $B \times B'$ such that $B, B' \in A$.)

Appendix II

On the Foundations of Probability

II.1. FOUR DIFFERENT METHODS FOR
INVESTIGATING PROBABILITIES

The classical conceptions of probability emerged in the seventeenth century as byproducts of the discussion on issues related to games of chance.[1] The question that was raised concerned the betting rates appropriate to different gambling situations. At first blush, the question seemed to be solvable by simple calculations. However, when different writers tried to give a comprehensive answer to the question, it became clear that such a solution could be attempted in different ways and that the relation between them was not self-evident.

a. The first type of answer was the following: If you want to inquire into the future results of, for example, the throw of a pair of dice, you should inquire into the physical aspects of the dice-throwing process.

b. Second, it was maintained that a prudent gambler should try to

---

[1] The earliest contributors to the modern theory of probability were probably Pascal, Fermat, and Huygens. In 1657, the latter wrote a treatise on "Rationacination in the Game of Dice," in which he gave the first definition of mathematical expectation. In 1713, Bernoulli published his famous "Ars Conjectandi," where he explicitly stated the additivity requirement for probabilities. De Moivre published in 1718 his "Doctrine of Chance," and here statistical independence and conditional probability were explicitly defined for the first time. In 1763, he published his brief "Essay Towards Solving a Problem in the Doctrine of Chances," where the famous Bayes theorem was stated. Then, in 1812, with his "Analytic Theory of Probabilities," Laplace summed up the basis for the modern theory of probability, stated his view on determinism, and concluded that probability must arise because of ignorance. For a book on the subject, see Todhunter, *The History of the Mathematical Theory of Probability*, Chelsea (New York, 1949). Original publication 1865.

224

gather information before gambling. For example, before gambling on the future results of tossing a coin, the gambler should toss the same coin a large number of times and use the frequencies of the types of results as empirical data that should enable him to determine his betting rates.

c. Next it was suggested that questions having to do with betting rates could be answered only by introducing special symmetry or invariance assumptions. For example, it was suggested that if we are ignorant with respect to the results of throwing a pair of dice, and if by observing the dice we do not detect any asymmetry, we should be led to believe that all of the possible results are equally likely. Therefore, if there are n possibilities and the prize is 1 dollar, we should be willing to gamble $1/n$ dollars on any one of the possibilities.

d. Finally, those who were dissatisfied with all of the attempts we have mentioned claimed that no betting rate could be ultimately justified in and of itself. The only thing a prudent gambler should do is to devise a coherent system of betting rates. But, above and beyond coherence, the argument went, there is nowhere to go. Betting rates are essentially idiosyncratic.

For lack of better terms, let us call the views physicalist, frequentist, logicist, and personalist, respectively. These views still form the major divisions among writers on the subject.

What makes the argument between proponents of these four views very confusing is that the views are all well motivated, but each points to a very different direction in the search for explanation and guidance: The first points to physics, the second to statistical research, the third to logic and mathematics, and the fourth points to normative theories of decision making. Now, some 300 years after the beginning of modern probability theory, it is fair to say that the research in all of these directions has proven fruitful.

a. Statistical physics was developed by Maxwell, Boltzmann, and Gibbs,[2] especially to explain why some physical systems have a

---

[2] Boltzmann's writings were collected by F. Hasenoehrl (Leipzig, 1909). Most of Maxwell's papers can be found in *The Scientific Work of J. C. Maxwell*, ed. W. D. Niven, Cambridge Univ. Press (Cambridge, 1890). Gibbs's work of interest was collected in *The Scientific Work of J. W. Gibbs*, eds. Bumstead and Van Name (N. Y. 1906).

disposition to assume certain types of states with constant probability. Although this research applies only to very limited types of actual systems, Hopf, one of the leading researchers in the field, actually tried to apply statistical physical methods to the case of a toss of a well-balanced coin.[3] He tried to prove that, in such a case, the probabilities of obtaining either heads or tails are indeed 0.5. It is too early to tell just how far this line of research can be developed, but it is a very lively discipline that attracts many mathematicians in many countries.

b. The "empiricist" or frequentist view was pioneered by Venn and Peirce in the nineteenth century. It developed into what is now called classical statistics through the work of von Mises, Neyman, Pearson, Fisher, Cramer, Reichenbach, and, most importantly, Kolmogorov.[4] These mathematicians and statisticians developed a very detailed theory of how to gather information in a reliable and efficient manner. Nowadays, the theory is applied almost everywhere and guides policy makers in virtually every field.

c. The mathematical concept of symmetry, the basis for the logicist or the symmetricist view, as we shall sometimes refer to it, was developed in various fields of mathematics and became one of the central concepts of twentieth-century mathematics. In its most abstract form, this idea was developed by Klein and Weyl in their pioneering work on group theory. [5] Later, however, the basic idea became so pervasive that, given any function (e.g., a probability function) defined on any set of objects (e.g., a σ-algebra of events), people began to look for some principles of invariance that would define the sense in which the function was symmetric.

The application of the concept of symmetry to probability was developed by J. Bernoulli, Bayes, and Laplace. They argued that, when we are presented with an exhaustive set of mutually exclusive

---

[3] See Hopf, *Ergodentheorie*, Springer-Verlag (Berlin, 1937) and also "On Causality, Statistics and Probability," *J. Math Phys.* 17 (1934), pp. 51–102.

[4] Let us mention only a few of the relevant titles: Cramer, *Mathematical Methods of Statistics* (Princeton, 1951); R. A. Fisher, *Statistical Methods and Scientific Inference*, Oliver and Boyed (Edinburgh, 1956); J. Neyman, *Lectures and Conference on Mathematical Statistics and Probability* Dept. of Agriculture (Washington, 1953); and of course Kolmogorov, *Foundations of Probability* (Chelsea, N.Y. 1956).

[5] For an introductory text, see Rotman, *Introduction to Group Theory*, Allyn and Bacon (Boston, 1984).

events, all of which are "equally possible," we should attribute to them equal probabilities. This argument, which is known as the law of insufficient reason, defines a concept of symmetry that is applicable to probability theory. Another related notion of symmetry can be defined for a sequence of events. The probabilities of the events in such a sequence will be symmetrical (or exchangeable) if they are invariant with respect to the order of the events in the sequence. The abstract mathematical theory that develops these concepts is measure theory. Lebesgue, Kolmogorov, and Doob[6] were among the principal contributors to measure theory. The mathematical theory of probability was applied to the question of the correct choice of betting rates by Keynes,[7] Jeffreys,[8] and Carnap,[9] who attempted to show how the adoption of special symmetry assumptions leads to a unique choice of a probability function.[10]

d. The application of the theory of decision to probability was made first by Ramsey and de Finetti,[11] the latter being the most important personalist of our time. Another important figure who worked on related issues is Carnap. Savage,[12] who took his cue from de Finetti, developed a personalist statistics that is arguably as powerful as the classical theory, and Jeffrey also

---

[6] See Halmos, *Measure Theory*, Van Nostrand (New York, 1950).

[7] See Keynes, *A Treatise on Probability*, MacMillan (London, 1921). Keynes credits many of his ideas concerning the principle of indifference to von Kries [see von Kries, *Die Principien der Wahrscheinlichkeitsrechnung* (Freiburg, 1886)]. In particular, Keynes was impressed with von Kries's critical remarks. (See ibid, chs. 4 and 5).

[8] See Jeffreys *Scientific Inference*, Cambridge Univ. Press (Cambridge, 1973).

[9] See Carnap, *Foundations of Probability*, Chicago Univ. Press (Chicago, 1950).

[10] The term logicism was coined by writers who believed that probability is a logical relation between a statement (expressing a hypothesis) and a set of statements (describing the available evidence). In practice, though, this relation is obtained by "translating" the rules of conditional probabilities. Now, those who support the claim that a logical relation holds between the evidence and the hypothesis have to show that the conditional probabilities are unique. The uniqueness is usually accomplished by special symmetry assumptions (for example, by assuming indifference principles of various kinds). That is why we believe that the main thesis of the logicists is that such assumptions are indeed necessary.

[11] The most comprehensive book de Finetti wrote is *Theory of Probability*, Wiley (New York, 1972).

[12] See Savage, *The Foundations of Statistics*, Wiley (New York, 1954).

developed a radical probabilist point of view in the spirit of de Finetti.[13]

Thus far we have presented a discussion on the foundations of probability theory via the issue of the choice of appropriate betting rates. We described a situation with a wide range of applications in which a prudent gambler, who tries to motivate his choice of betting rates in a particular gambling situation,[14] finds out that there are no less than four different methods that he could use. What makes matters even more complex is that these methods are not even necessarily incompatible! In fact, as we shall see shortly, when people tried to pair the physicalist, frequentist, logicist, and personalist insights, only one combination did not lead to a coherent view: physicalism and personalism. Therefore, it seems hasty to even talk about four competing foundational schools!

In particular, we should resist the idea that behind these four insights there are four different definitions of probability. Such a thought does indeed cross one's mind. Physical probability could be defined as a property of physical systems. Empirical probability would be a formal feature of sequences of events. Symmetricist probability could be taken to be a logical relation between propositions, or a hybrid of such relations and a background theory. And, finally, personalist probability would measure the intensity of the beliefs of the agent.

There are two reasons why we resist the idea that there are four definitions of probability. The first is that, insofar as the four positions we described are not necessarily incompatible, there is no reason to introduce such incompatibility by positing more definitions than necessary. But, more to the point, neither the idea of frequentist probability nor the idea of logical probability seems to withstand scrutiny. To be sure, the original insights in both cases are extremely important, but, as we shall see shortly, both of these ideas are consistent with both physicalism and personalism. Therefore, positing four different types of probabilities seems unnecessary; we can make do with two types of

---

[13] See de Finetti, *The Logic of Decision.*

[14] The theory of epistemic decision making shows how to extend the concept of gambling to a whole variety of situations where an agent must decide which proposition to accept on the basis of his current knowledge. The main idea is to introduce a concept of epistemic value that will function as the payoff. See Levi, *Gambling with Truth*, Knopf (New York, 1967) for a theory that develops these ideas.

probabilities: one that stands for objective, external, and real properties of physical systems; and another that denotes degrees of belief in propositions, that is, subjective probabilities.

Remarks

a. The connection between physicalism and empiricism was investigated in depth primarily by mathematicians and physicists. The mathematical theory that tries to account for the connection between frequency phenomena and physical properties is called ergodic theory.[15] We return to this theory often throughout the book. A philosophical defense of this perspective was attempted by von Plato.[16]

b. The possibility of making justifiable symmetry assumptions is researched in virtually every field of theoretical physics. When there is a physical reason for a general symmetry assumption concerning a space of events, this reason may become a bridge between the physicalist and the symmetricist views concerning the probability of those events.[17]

c. De Finetti and Savage showed how it is possible to present many of the classical empirical techniques as special cases of a more general personalist point of view. Consequently, they claimed that the idea of collecting data is not in and of itself incompatible with personalism. We shall return to this subject in Section II.5 of this appendix.

d. The introduction of symmetry assumptions into personalism is, again, a well-trodden path. Both Carnap and Jeffreys, for example, presented symmetry assumptions as extended coherence requirements. Those who hold such views are called *necessitarians*. Jaynes, whose views were discussed in the first chapter, is another writer in this tradition.

e. The only incompatible pair, as we have remarked, is physicalism and personalism. If physical systems have physical probabilistic disposi-

---

[15] For an introductory text, see Walters, *Introduction to Ergodic Theory*, Springer-Verlag (New York, 1976).

[16] See von Plato, "The Significance of the Ergodic Decomposition of Stationary Measures for the Interpretation of Probability," *Synthese* 53 (1982).

[17] Boltzmann, as we have seen in Chapter 1 of this book, introduced the assumption that the probability of obtaining a pair of colliding molecules with prescribed velocities is symmetric with respect to time.

tions, there is nothing idiosyncratic or agent-dependent in these probabilities. That is why physicalism is widely regarded as an objectivist view of probabilities. Personalists who insist that probability depends on the state of knowledge of the agent are known as subjectivists. There is, however, a long tradition of calling both the physicalist and the empiricist views objectivist while calling the symmetricist and personalist views subjectivist. This tradition was formed for a good reason. Most of the classical statisticians talk about their job as one of estimating fixed but unknown parameters, which justifies calling them objectivists. On the other side, in this century the leading writers who stressed the role of symmetry in probabilistic reasoning were revisionist–personalists who thought that they were extending the sphere of coherence requirements.[18] This division makes the subjectivist–objectivist distinction dominant in the discussion of the foundations of probability.

f. Finally, there are writers who are dualists. Like the objectivists, they believe that in some cases physical systems have objective probabilistic properties. In those cases, the dualists recommend that people who know these probabilities use them as their personal probabilities when making decisions. However, the dualists also warn us not to forget that we always need personal probabilities to make decisions. In cases where these probabilities cannot be guided by knowledge of objective probabilities, the dualist maintains that personal probabilities are to be regarded as personal opinions. Carnap, for example, was a dualist. A different dualist theory also was developed more recently by Levi.[19]

g. Let us close this set of remarks by mentioning that the logicist and empiricist insights themselves are consistent with each another. The introduction of symmetry assumptions is a tool that is used constantly by classical statisticians, who regard such

---

[18] Let us not forget, however, that the idea of using symmetry considerations preceded any clear formulation of the personalist point of view. This is particularly true of the writing of Pascal, Fermat, and Huygens. More generally, the relation between the logicist and the subjectivist approaches is far from clear. The tension between the two views was discussed in Chapter 1 in relation to the work of E. T. Jaynes.

[19] See, for example, Levi, *The Enterprise of Knowledge*, MIT Press (Cambridge, MA, 1980), chs. 11 and 12.

assumptions as a part of the theoretical background of the experimental setup.[20]

This, then, is the spectrum of views on probabilities. There are many issues, many different views, and many possible ways to synthesize the existing views, but there is just one dividing line: There are those who think that probabilities are external, real, and fixed, and those who think that probabilities are expressions of personal attitudes resulting from ignorance. However, as we have remarked, neither the empiricist nor the logicist is required to make up their mind on the objectivist–subjectivist issue. Their views are consistent with both objectivism and subjectivism. It should be noted, though, that most of the writers on the foundations of probabilities did in fact make up their minds with respect to their allegiance to either subjectivism, objectivism, or dualism. Hence, bearing in mind all of the complexities that we have alluded to, we can talk about subjectivism and objectivism as the two major foundational camps.

Let us now make a remark that may appear somewhat trivial: We have characterized as physicalists those who believe that investigating the physical makeup of systems is the best method for learning about the probabilities with which these systems assume certain states. We also said that the discipline that investigates these questions is statistical physics. This, however, should not be taken to mean that statistical physicists are necessarily physicalists with respect to the probabilities that appear in SM. First, there are different reasons for studying SM that have little to do with the foundations of probabilities.[21] Second, and more importantly, statistical physicists hold very different views concerning the foundations of probability. Some of them are personalists, others are empiricists and logicists who allied themselves with subjectivism, and, of course, many physicists are empiricist–symmetricists who accept the objectivist point of view.[22] This plenum of views makes SM an interesting battleground.

---

[20] For an attempt to integrate the logicist and the frequentist approaches, see Kyburg, *The Logical Foundations of Statistical Inference*, Reidel (New York, 1974).

[21] One of the original reasons for studying SM was to justify atomism in physics. The well-known controversy between Boltzmann and Mach revolved around this issue.

[22] Einstein, for example, thought that physical probability should be reduced to a measure of frequency.

The reason why the distinction between objectivists and subjectivists divides the community of physicists is a profound one. SM poses the opposition between subjectivism and objectivism in a way that makes both subjectivism and objectivism very well motivated. On the one hand, the intuition that the probabilities of physical systems can be construed as physical parameters is a strong one. But, on the other hand, physicists who subscribe to the usual Newtonian determinism will not find it easy to admit objective probabilities into classical physics. (These physicists are not always aware of the fact that there is also a tradition of trying to define objective probabilities in a way that is indifferent to the issue of determinism.) [23]

Therefore, in our opinion, an assessment of how well the two foundational camps fare with respect to SM is interesting. In particular, we were interested in invstigating whether subjectivism is a coherent view with respect to SM. This question is one of the central themes of this book.

We shall dedicate the next four sections of this appendix to a systematic exposition of the subjectivist's point of view.

II.2. SUBJECTIVIST FOUNDATIONS OF PROBABILITY

Subjectivism concerning the probabilities that appear in the formulation of a scientific theory is primarily an eliminative program. The formalism of the theory conveys the impression that the probabilities have to be given objectivist interpretation. The subjectivist, instead, has to show that this need not be the case. He must argue that we do not need to give the probabilities in question an objective (i.e., external, real) interpretation, but that they can instead be treated as propositional attitudes – as degrees of belief. These degrees of belief, the subjectivist maintains, are formulated because of ignorance with respect to the precise data needed for an exact prediction.

Formally, the degrees of belief of an agent X in propositions that describe events[24] can be represented as a function that assigns to each proposition A a real number $p \in [0:1]$. The degree of belief p that an

---

[23] Venn was one of the pioneers of this tradition. A contemporary defender of this program was Ernest Nagel; see, e.g., *The Structure of Science*, Harcourt, Brace and World (New York, 1961). There is also a tradition of writers who adhere to subjectivism without appealing to determinism (de Finetti, for example).

[24] The set of descriptions is usually assumed to be closed under countable unions and intersections, i.e., the descriptions form a σ-algebra.

agent has in A (henceforth $\deg_x(A) = p$) is said to be revealed in his betting behavior – p dollars is the stake he or she would bet on A being the case, where the prize is 1 dollar. The constraints on the choice of degrees of belief are said to be derived from the constraints on a coherent betting behavior. In proofs called Dutch Book arguments, the subjectivists show that an incoherent system of degrees of belief that does not satisfy these requirements will result in a sure loss.

The restrictions on a coherent degree of belief function are the following:

Constraint 1. *The degree of belief in what you know to be the case ought to be 1.*

Constraint 2. *If* A *and* B *are descriptions of events such that you know that* A *and* B *cannot be the case together, then*

$$deg_X (A \ or \ B) = deg_X (A) + deg_X (B).$$

Another important concept that should be introduced is $\deg_x(A/B)$, the conditional probability of A given B. The subjectivist thinks of the conditional $\deg_x(A/B)$ as the degree of belief that you would have in A if you knew that B were the case. The value assigned to the conditional $\deg_x(A/B)$ expresses the amount that you are willing to bet on the proposition "A and B" when you know in advance that the bet will be called off if B turns out to be false.[25] Therefore, the choice of conditional degrees of belief is constrained in the following way:

Constraint 3. $deg_X(A/B)deg_X(B) = deg_X(A \ and \ B)$.

The above equation describes the way in which conditional and nonconditional probabilities ought to constrain each other. More generally, all of the other constraints are to be interpreted in this way. Subjectivists do not prescribe holding any particular degree of belief unconditionally but demand only coherence, a property of the whole body of partial beliefs.

Let $B_i, 0 \le i \le n$ be a partition of descriptions of events. This means that (i) for some i, B_i must be the case, and (ii) if $i \ne j$, B_i and B_j cannot

---

[25] For his treatment of conditional probabilities in terms of "called off" bets, see de Finetti, *Probability, Induction and Statistics*, Wiley (New York, 1972), pp. 14–16.

be the case together. One of the consequences of the three constraints that we have introduced is known as Bayes's theorem. If C is a description of an event, then

$$\deg_X(B_i/C) = \frac{\deg_X(C/B_i)}{\sum_{i=1}^{n} \deg_X(C/B_i)\deg_X(B_i)}$$

If we think of $\deg_X(C/B_i)$, $\deg_X(B_i)$ as being fixed, Bayes's theorem shows us how to calculate $\deg_X(B_i/C)$ from $\deg_X(C/B_i)$ and $\deg_X(B_i)$. Because of that, it is often claimed that Bayes's theorem shows how to update $\deg_X(B_i)$ once new data C become known. As we remarked, we are not forced to proceed in this direction, because, strictly speaking, the above formula shows only how $\deg_X(C/B_i)$, $\deg_X(B_i/C)$, and $\deg_X(B_i)$ ought to constrain each other.

The above constraints are accepted by subjectivists across the board. These constraints do not single out a unique rational probability and, according to the subjectivist, they should not be expected to do so. Subjectivists maintain that other constraints are self-imposed and hence arbitrary.

II.3. SUBJECTIVISM AS AN ELIMINATIVIST METHODOLOGY

Now we can formulate the subjectivist elimination program, which is an attempt to show that the probabilities in the scientific context in question can be represented as partial beliefs without there being a loss of explanatory power. More specifically, subjectivists often advance arguments of the following types (this is a tentative list):

1. They will argue that probabilities arise in different contexts for precisely the same reason, namely, because of ignorance. Without ignorance, the whole point of introducing probabilities disappears.
2. Because probabilities arise for the same reason, the subjectivist argues, the formal properties of probabilities are general in character and the same in all situations. Therefore, the subjectivist accepts very few constraints on the assignments of probabilities so that his concept of probability will remain as general as possible. In particular, most subjectivists refuse to impose on their probabilities the σ-additivity constraint. [A probability P is σ-additive if and only if, given a countable partition $\{A_n\}_{n=1}^{\infty}$, $\sum_{n=1}^{\infty} P(A_n) = P(\cup_{n=1}^{\infty} A_n)$.]

3. Probabilities cannot be definable in terms of nonprobabilistic notions, because they are not mathematically equivalent to non-probabilistic notions; nor can probabilistic judgments be deduced from nonprobabilistic judgments.
4. There is no way to connect probabilities to empirical findings directly; i.e., there is no physical magnitude that expresses a "correct" probabilistic assignment.

When these points are argued, the subjectivists claim that probabilities are "figures of speech," and that they are "in our heads," and that they are not external and real; that is, they are eliminated. However, we should keep in mind that there are senses of elimination that are different from that which the subjectivist developed. We should not confuse the triumphant rhetoric of the subjectivist with the similar rhetoric voiced by the proponents of other eliminative programs.

In particular, we should not confuse subjectivism with an elimination in the sense of Hilbert, Quine, or Field. These three authors advanced the following programs of elimination:

1. From Hilbert's attempts to eliminate infinite totalities from mathematics, we may deduce the following elimination method: If you show that any proof that uses a concept can be replaced with a proof that does not use it, you have eliminated the concept.[26]
2. Quine's method of eliminating an entity consists in a demonstration that it is definable from other concepts. That is, it is a proof that there is no need to introduce the entity explicitly into the ontology of the theory in question.[27]
3. In his *Science Without Numbers*, Field used the following elimina-tive methodology: To eliminate a theory is to show that any truth

[26] In Hilbert's finitist program, the proofs themselves have to be finite.
[27] Quine's idea can be given a more precise formulation when the notion of a defini-tional extension of a theory is given. If T is a theory and t is a term that can be defined by using only the terms of the language of T and logical symbols, then when t is added to the language of T the result is a definitional extension of T. When a concept or an entity is presented in this manner, by using the terminology of Quine we have elimi-nated the need for a separate ontological commitment concerning the entity or the concept in question. From that point on we may regard the term denoting the entity or the concept to be a shorthand, as a mere linguistic convenience. (The proof that the elimination is possible, though, may be highly nontrivial.)

that the theory entails can be entailed without the theory. That is, to eliminate a theory you show that the theory does not give rise to new truths.[28]

The subjectivist method of eliminating objective probabilistic concepts is to be contrasted with the above-mentioned senses of "ontological elimination."[29] In fact, when the subjectivist represents the concept of probability as a propositional attitude of partial belief, the result is not an elimination of the concept in any of the above senses. Propositional attitudes exist as mental concepts that give rise to certain dispositions (for example, dispositions to gamble in a particular manner). However, subjectivized probabilities no longer belong to the class of ordinary theoretical terms. They do not refer to external objects, and statements about them are neither true nor false. According to the subjectivist, the sentence "the probability of p is x" merely expresses our own attitude. Therefore, the idea that in SM, for example, the probabilities are to be interpreted in this manner is an idea with far-reaching consequences.

We believe that the subjectivist approach to probabilities should be discussed in a more general way. One should try to look at the way in which de Finetti treated probabilities and see whether other concepts can and should be treated in the same manner. One of the reasons why the general methodological character of subjectivism has been overlooked has to do with the fact that de Finetti, the most important subjectivist of our time, treated methodological issues with a hefty amount of impatience, which influenced the writings of those who followed him. This is unfortunate, because there are many interesting issues that reveal themselves only when subjectivism is presented as a general thesis. One of these interesting issues – which de Finetti addressed only briefly – is the connection between de Finetti's elimination of prob-

---

[28] Field wrote his argument in *Science Without Numbers* (Princeton, 1980). Note that the notion of semantic entailment is not equivalent to the syntactic notion of a proof. A sentence is semantically entailed by a theory if it is true in all of the models of the theory. A sentence that is entailed by a theory may not be provable from the theory.

[29] In Chapter 3 we surveyed the construction of the Haar measure. The Haar measure is obtained as the unique measure that satisfies a list of operational constraints, none of which mention any probabilistic concepts. This construction therefore can be viewed as an elimination of probabilities in the Quinean sense of elimination. As we saw, such a construction is possible only in very degenerate cases.

abilities and Hume's elimination of causation.[30] By using the terminology we have developed, we may say that Hume's is a program for the subjectivist elimination of causation. Hume attempted to show that, when causation is treated as a psychological concept or as an association that the mind makes between ideas, the resulting theory does not suffer any loss of explanatory power. Once he defined causation that way, Hume thought that he had eliminated causation. This sense of elimination, like de Finetti's, cannot be captured by the competing concepts of elimination that we mentioned earlier. More generally, an abstract discussion of subjectivism may suggest other candidates to be eliminated in a subjectivist manner (one candidate, as we have seen, is the concept of approximation). Let us, however, abandon the general discussion for the time being and examine instead an important example of the practice of subjectivist elimination, namely, de Finetti's treatment of statistical estimation.

II.4. SUBJECTIVIST FORMULATIONS OF STATISTICAL CONCEPTS

Let us begin our discussion with a case that both objectivists and subjectivists regard as an unproblematic statistical estimation. Suppose that we are presented with an urn that contains black and white balls. To estimate the percentage of black balls in the urn, we may draw and replace n balls from the urn and regard the percentage of black balls in the sample as an estimate of the percentage of black balls in the urn. In this case, the subjectivist will agree that, speaking about the percentage of black balls in the urn as a fixed and unknown probability, what we are estimating can be treated as a harmless manner of speech.[31]

Matters become more complex when objectivists talk about estimating the unknown but fixed probability that a coin will land heads up on a toss. For the objectivist, tossing the coin a large number of times and using the percentage of heads in the sample as an estimate of the probability of heads is perfectly analogous to the case of the urn. The subjectivist, on the other hand, will regard this reasoning as the case of an analogy stretched too far. "There is nothing to be estimated!" says the subjectivist. "It is as if we were to describe a wireless phone as a

---

[30] For a brief remark of his on the subject, see de Finetti, *Probability, Statistics and Induction*, Wiley (New York, 1972), p. 182.

[31] See ibid, pp. 229–230.

phone with an invisible wire!" "But, be that as it may," says the objectivist, "how else can we describe what we do when we toss the coin a large number of times and use the information obtained from the sample to fix our degrees of belief?" To meet this challenge, the subjectivist must reconstruct or reformulate in his own terms a general theory of statistical estimation.

The reconstruction begins with the most basic definitions. The objectivist describes the series of coin tosses as a sequence of repetitions of the same trial. The subjectivist will speak about a series of unknown events that we choose to regard as a sequence. The objectivist will define the notion of the frequency of heads as a property that is somehow connected to the "chance setup" that produced the sequence. The subjectivist, on the other hand, will ask, Exactly how many of the events were tosses that yielded heads?[32] The objectivist will talk about the limit of the frequencies among the members of a sequence when its length approaches infinity. For the subjectivist, this mode of speech is unacceptable, and he will make do with asking about the percentage of heads in longer and longer sequences. Finally, the objectivist will state that one may use the frequencies in a long sequence of events as estimates of the unknown probabilities, but the subjectivist will claim that we can only hope to estimate unknown long-run frequencies. In other words, we can only ask whether continuing the tossing of the coin would result in a series of tosses that yielded heads in the same proportion as before.[33]

There is, however, at least one more issue that the subjectivist must confront. The objectivist refers to the sequences of tosses as independent trials, by which he means that the probability of a sequence of tosses is the product of the probabilities of each of its members. Such probabilities are called Bernoullian. The objectivist will justify his claim that the probabilities of the coin tosses are Bernoullian by saying that the mechanism that produces the coin tosses does not have a memory, and so the result of one toss cannot influence the result of the next toss. For the subjectivist, on the other hand, it is our memory and not the memory of the mechanism that is the issue. Indeed, it seems that the

[32] The subjectivists, as we shall see shortly, developed their own way of treating the notion of frequencies. To this end, de Finetti proved a representation theorem that we shall discuss in the following pages.

[33] For his account of these issues, see de Finetti, "Foresight – Its Logical Laws, Its Subjective Sources" in *Studies in Subjective Probability*, ed. Kyburg and Smokler, Krieger (New York, 1980), pp. 71–78. See also "Probability . . . ," ibid, pp. 229–230.

knowledge that the last nine tosses yielded heads should influence our degrees of belief that the next toss will yield heads as well. At this point the difference between the two views ceases to be one that may be resolved by redefining the terms and becomes a substantive disagreement. It is also at this point that the subjectivist has to face another related issue. There is a series of important theorems that tie together frequencies and probabilities, and these theorems apply primarily to Bernoullian probabilities. These theorems are known as laws of large numbers.

Let P be a probability defined on S, the space of infinite sequences of heads and tails. P assigns probabilities to the σ-algebra of the cylindrical sets of S. (These sets are all of the countable unions, intersections, and complements of sets of sequences whose ith coordinate is stipulated to have a particular value.) A probability measure P is a Bernoulli probability if, for all n,

$$P(\{w_0, w_1, \ldots \mid w_0 = \text{heads and } w_1 = \text{heads and} \ldots \text{ and } w_n = \text{heads}\}) = q^n,$$

where q is the probability of heads on a toss. Let $w_0, w_1, \ldots$ be a member of S. The frequency of heads in $w_0, w_1, \ldots$ can be defined in the following way:

$$\text{freq}_H(w_0, w_1, \ldots) = \lim_{n \to \infty} \frac{1}{n} \sum_{i=1}^{n} X_H(w_i)$$

where $X_H(w_i) = 1$ if $w_i = \text{heads}$ and $X_H(w_i) = 0$ otherwise. Now we can formulate the strong law of large numbers.

Theorem (Borel 1909).[34] *If P is a Bernoulli probability on the space of sequences of heads and tails, then (a) the probability that P assigns to the set of infinite sequences that have heads with frequency q is 1; and (b) the probability that P assigns to the set of sequences that have an initial segment where heads appears with deviant frequencies becomes lower in proportion to the deviance and the length of the initial segment.*

[34] In fact, Borel's proof applies only to the case where $p = 0.5$. Cantelli generalized the proof for arbitrary probabilities; see Borel, *Probability and Certainty*, Walker and Co. (New York, 1950), p. 27. For a very illuminating discussion of the law of large numbers and its relation to the subjectivist point of view, see von Plato, "The Significance of . . ." (see note 14), pp. 420–425. For a standard formulation of the law of large numbers and a proof, see, e.g., Halmos, *Measure Theory*, ibid, p. 201.

Because of (b), the observed frequencies have a clear relevance to the objectivist. He can say that, with a high probability, the observed frequency in an initial segment of a sequence will not be significantly different from the unknown but fixed value of p.

The full subjectivist response to this argument is an issue that was discussed at length. The main issue concerns whether there are instances, other than the Bernoulli case, where analogies of the laws of large numbers can be proven. There is also the question of whether the objectivist assumptions are indeed necessary for the applicability of the laws of large numbers. Here, de Finetti himself made a substantial contribution when he proved that exchangeable probabilities, that is, those probabilities that disregard order, can be written as weighted averages of Bernoulli measures. As a result, subjectivists can apply the law of large numbers in those cases where exchangeability can be assumed. We shall come back to these issues in later sections.[35]

II.5. STATISTICAL EXPLANATION

The second context in which subjectivists will attempt to eliminate objective probabilities is statistical explanation. This context is very important to our concerns, because SM offers statistical explanations to many truths that are otherwise left unexplained.

The modern treatment of statistical explanation begins with a formative analogy. It is maintained that inductive and deductive nomological explanations are essentially analogous. A deductive nomological explanation, as the reader may recall, is a derivation of the form

DN all A are B
D a is an A
C a is a B.

That a is a B is explained when it is derived from a general law-like sentence DN and particular data D on a.

Similarly, the argument goes, a statistical explanation is also a derivation.

---

[35] For the original proof, see de Finetti, *Studies in Subjective Probabilities*, ibid, pp. 78–100.

S   *A* are mostly *B* (or, *A* are very likely to be *B*)
D   *a* is an *A*
C   *a* is a *B*.

Here, too, the analogy goes, that *a* is a *B* is explained when we show that it is an instance of a general statement *S* that predicted it with a high probability.[36]

The analogy was shown to be very problematic. The following is a summary of some of the points that appear in the vast literature on the subject.

1. The DN explanations are based on valid inferences. That is, they are based on principles that allow us to move only from true premises to true conclusions. By its very definition, however, the statistical analog permits cases where the premises are true and the conclusions are false. Why, then, are statistical explanations valid? And, if the concept of validity itself needs to be generalized, what could such a generalization be?

2. Another disanalogy with the deductive case is that if a sentence *S* follows deductively from a set of sentences *G*, it follows from every set *G'* that includes *G*. This feature is known as *monotonicity*. Inductive inferences are not monotonic. It is because of this peculiarity that Hempel argued that inductive inferences only state the existence of a relation between the premises and the conclusions.[37] This anomaly can be overcome if inductive inferences are required to be made from the maximal set of relevant premises. This is known as the *total knowledge requirement*.[38] There is no need for a similar requirement in the deductive case.

---

[36] There are many formulations of the statistical premise *S*. For Hempel, *S* is of the form "the statistical probability of an *F* being a *G* is nearly 1." The expression statistical probability is used in a somewhat vague manner; see *Aspects of Scientific Explanation*, Free Press (New York, 1965), p. 55. For a criticism of Hempel's view, see, e.g., Salmon, "Scientific Explanation" in *The Nature and Function of Scientific Explanation*, ed. Colodny, Pittsburgh Univ. Press (Pennsylvania, 1970). In Chapter 4, another nonprobabilistic formulation of statistical explanation is introduced, where *S* is of the form "The generic members of *A* are *B*." It is important to note that the discussion in this section applies essentially to all of these versions.

[37] See Hempel, "Aspects . . . ," for a discussion on all of these issues.

[38] This requirement was formulated by Carnap in *The Logical Foundations of Probability*, Chicago Univ. Press (Chicago, 1950), p. 211.

The total knowledge requirement can block the undesirable cases because when we know that:

$$\text{most } A \text{ are } B$$

and

$$a \text{ is an } A$$

we cannot automatically infer that

$$a \text{ is a } B$$

if, in addition, we know that

$$a \text{ is a } C.$$

Notice that, without such a requirement, we would have had a straightforward contradiction if, in addition, we believed that

$$\text{most } C \text{ are not } B.$$

Essentially, this is what is known in the literature as the *reference class problem*.[39]

3. In the deductive case, given a set of premises, the conclusion is either acceptable or not. When dealing with induction, we always select from a set of rival hypotheses. This strengthens Hempel's conclusion that an inductive inference is a complex relation. One of the factors that determines whether the relation holds is the set of potential conclusions.[40]

4. Some writers argue that the dependence of inductive inference on the context of inquiry goes even further to include auxiliary parameters, such as the demand for information and the degree of caution we wish to exercise. More generally, inductive acceptance involves complex decisions that are completely absent from the deductive case.[41]

5. Salmon argued[42] that the statistical premise S does not have to be

---

[39] See Hempel, "Inductive Inconsistencies," *Aspects*, ibid, pp. 53–79.
[40] On this point, see Levi, "Four Themes in Statistical Explanation" in *Causation and Decision, Belief Change and Statistics II*, eds. Harper and Skyrms, Kluwer (Roterdam and Boston, 1988), p. 207.
[41] Ibid, pp. 208–213.
[42] See Salmon, "Statistical Explanation," p. 177.

of the form "the probability that A are B is q," where q is close to 1. An example of this sort is the following: Suppose that we explain why X caught a cold by stating that he was adjacent to Y, who had a cold. When we give such an explanation, we do not necessarily believe that the probability of catching a cold from a person who stands near us is close to 1. However, we do believe that the probability of catching a cold without being near a person with a cold is much lower than the probability of catching a cold after having been near such a person. That is why the above argument is indeed explanatory.

All of this points to the fact that the analogy between DN explanations and statistical explanations is a weak one. On the basis of

$$S \quad A \text{ are mostly } B$$

and

$$D \quad a \text{ is an } A,$$

we conclude that

$$C \quad a \text{ is most likely to be } B.$$

And even that may be defeated in certain cases.

So far, we have not said much about the logical form of the statistical premises S, because we think that the problems posed by the disanalogy between the DN explanation and the statistical ones are largely independent of questions of the exact logical form of S. However, there are considerably different conceptions of what a statistical explanation is, depending on what S is taken to be.

For the subjectivist, the issue of statistical explanation is, finally, rather trivial. S is of the form $\deg_x(B/A) = q$, D is that a is an A, and the conclusion is that $\deg_x(a \text{ is a } B) = q$. To be sure, one should be aware of the reference class problem, but conceptually the fact that a new piece of knowledge about a may change our degree of belief concerning a is not at all problematic.

It is only when S is taken to be an objectivist chance statement that statistical inference becomes a serious issue. At least on the face of it, relative to different reference classes, different truths are being contemplated. Hence the objectivist will try to make sure that the conditions under which a system is attributed with a chance property are stringent enough to preclude the emergence of reference class problems. In SM it is usually possible to satisfy these conditions, because they are built into the descriptive apparatus of the theory, so to speak. One of the outcomes of the quasi-ergodic hypothesis is that we should not worry too much about either the reference class problem or issues having to do with direct inference in the context of SM.[43] The issue is addressed in Chapter 2.

[43] Von Plato writes: "Our conclusion is that probability as time average solves in special cases . . . the problem of reference class. Almost all samples turn out fair for the proposed criterion of homogeneity." See "Ergodic Theory and the Foundations of Probability" in *Causation, Chance and Credence*, vol. 1, eds. B. Skyrms and W. Harper, Kluwer (Roterdam and Boston, 1988), pp. 257–277.

Appendix III

Probability in Nonequilibrium Statistical Mechanics

III.1. IS THE PROBABILISTIC FRAMEWORK OF EQUILIBRIUM SM TOO RESRICTIVE?

The main objective of this book has been to analyze the concept of probability as it is defined and used in the context of statistical mechanics. Other topics that arise in the discussion on the foundations of statistical mechanics have been given much less attention and emphasis. A case in point is nonequilibrium statistical mechanics (henceforth NESM) in general and the issue of irreversility in particular. These issues are, no doubt, extremely important, so much so that, from certain points of view, most of the nontrivial aspects of statistical mechanics necessitate the introduction of the framework of NESM. What distinguishes this framework, some writers maintain, is the presence of time-dependent probabilities.

In this appendix we shall argue that the concept of probability in NESM is still ill understood. We maintain that those aspects of NESM that can be given a coherent and rigorous treatment can be discussed in a framework that is only a slight generalization of the framework of equilibrium statistical mechanics. Conversely, those aspects of NESM that transcend the equilibrium framework do not have, as of yet, secure foundations. The main object of our discussion will be the conception of NESM that was developed by the so-called Brussels school. We shall argue that the beautiful conceptual analysis of NESM that the members of this school developed do not yield, at present, a substantial generalization of the framework of equilibrium statistical mechanics.[1]

---

[1] The view of NESM that we shall present was discussed in Balescu, *Equilibrium and Non Equilibrium Statistical Mechanics*, Wiley (New York, 1975), and in the following articles by Dougherty: "Explaining Statistical Mechanics," *Stud. Hist. Phil. Sci.* 24 (5) and "Foundations of Non-Equilibrium Statistical Mechanics," *Phil. Trans. R. Soc. Lon.* (1994), p. 346.

A similar verdict applies in the case of Jaynes's conception of NESM as well.[2]

III.2. INTRODUCTION OF PROBABILITIES INTO THE
SM OF CLOSED SYSTEMS

Let us begin by reviewing once more some aspects of equilibrium statistical mechanics. In particular, let us recall how probabilities are introduced in this context. We shall begin by reviewing the simplest case, namely, the statistical mechanics of isolated systems.

Let M be an n-particle system. We shall say that M is isolated if no energy is gained or lost. Hence, as long as M remains isolated it is always in a fixed energy level. More generally, we require that all of the forces that operate on the system have been identified and that as long as the system remains isolated no new forces will be operative. Hence, once the state of the system is fixed, the subsequent development of the system is completely determined. In particular, when the system is observed at different times and found to be in the same state, the fact that the observation takes place at different times makes no difference. Therefore, there is no need to introduce the time of the observation as an additional variable. The time could have made a difference only if new forces, yet unidentified, present themselves at a certain point of time and, once operative, they affect the development of the system. However, when isolated systems are concerned, this possibility is ruled out explicitly. Hence, beyond the specification of the state of the system, no other parameter is needed for the determination of the evolution of the system. In particular, no mention needs to be made of the time of the observation.

This general rule applies in particular to the transition probabilities of a stationary system. If they are taken to be physical properties, these probabilities should also be determined as functions of the state of the system alone. In general we are looking for a way to determine $p(x_{t+\Delta t} \in A, x_t \in B)$, that is, the probability that M will be in an A-state at $t + \Delta t$ given that it was at a B-state at t. When stationary systems are concerned, the time t is not important. What is important is the time difference, Δt. In particular, without limiting the scope of the argument, we may set $t = 0$.

---

[2] For Jaynes's view of NESM, see his *Papers on Probability Statistics and Statistical Physics*, ibid, pp. 287–298.

From a mathematical point of view, a stationary system M is represented as the measure-preserving transformation $\langle S, B, m, T \rangle$, where, for every $A \in B$ $m(A) = m(T(A))$. The reason for the identification was presented in Chapter 2.[3] As the reader may recall, the argument was as follows. First, we construct a system $\langle S', B', m', H \rangle$, where S' is the set of states of M, B' is the set of measurable subsets of S', H is an abstract representation of the Hamiltonian as a 1–1 and onto measurable transformation $H : S' \to S'$, and m' is a measure on B. Next, let us define a new space $\mathbf{S} = \langle S, B, m, T \rangle$ in the following way. As the points of S let us take the trajectories of $\langle S', B', m', H \rangle$. That is, a point of S is a doubly infinite sequence $\{ \ldots H^{-1}s, s, Hs, H^2s, \ldots \}$, where $s \in S'$. As the set B of measurable events of S we shall take the σ-algebra generated by cylindrical sets of the form $\{ \{ \ldots, w_{-1}w_0w_1w_2, \ldots \}$ s.t. $w_i \in A \}$, where $A \in B'$. T is defined as the shift transformation, that is, $T(\{ \ldots, w_{-1}w_0w_1w_2, \ldots \}) = \{ \ldots, w_{-2}w_{-1}w_0w_1, \ldots \}$. T assigns to each trajectory another trajectory that has the same elements from S' arranged in the same consecutive order but parametrized differently. That is, the w_i-th element of s is the $w_{i\text{-}1\text{-th}}$ element of $T(s)$. The shift transformation can be interpreted as the "passage of time." If we think of the 0th element as the state of the trajectory in the present, then the shifted trajectory corresponds to a system that is identical to M in all respects but was observed one moment of time later. This interpretation makes it clear why when $\langle S', B', m', H \rangle$ is stationary the corresponding shift $\langle S, B, m, T \rangle$ is measure-preserving. Indeed, in Chapter 2 we proved a theorem to that effect.[4]

III.3. SYSTEMS THAT ARE IN A STATE OF EQUILIBRIUM WITH RESPECT TO THEIR SURROUNDINGS

Our next step is to remind the reader that all of the results that obtain for isolated systems remain true for an entirely different set of cases, namely, to systems that are in a state of equilibrium with their surroundings. This argument is important because real systems are not completely isolated from their surroundings. Therefore, if SM were to apply only to completely isolated systems, it would not be of much use.

---

[3] See pp. 115–116. For a similar argument, see Peterson, ibid, pp. 6–7.
[4] Ibid.

Another point worth mentioning is that the idea of extending the methods of SM to "almost isolated" systems is completely hopeless, because systems with many particles are not structurally stable.[5] Even the addition or subtraction of a modest amount of energy may change the character of the trajectories of the system dramatically. This means that the fact that even if a system is "almost isolated" we may not use any approximation techniques to describe the behavior of the system. Therefore, the fact that the methods of SM apply in the case of systems that are in a state of equilibrium with respect to their surroundings is very important indeed; if they did not, SM would not have much empirical content.

A system that is in a state of equilibrium with respect to its environment interacts with the environment freely; it exchanges energy with its surroundings all of the time. Hence, such a system cannot be in a state of constant energy. However, after a certain relaxation time, we expect the situation to stabilize; a certain point will be reached where the flow of energy from the system to its environment will be the same, on the average, as the flow from the environment to the system. In such a case, the temperature of the system will equal that of its surrounding. Once this state is reached we might say that, although the energy of the system is not fixed, the expectation of the energy is. This means that if $E1, E2, \ldots$ are the possible energy levels, the expectation $\Sigma E_i p(E_i) = E$ is fixed. When the system is made of many particles, the expected value is also the most probable value. In such a case, there will be a single energy level E that is vastly more probable than all of the others. In addition, the fluctuations from E will be relatively small.

Let us take the argument further and assume, as it is customary to do in SM, that we may treat macroscopic systems as if they were made of infinitely many particles, that is, let us assume that we may exchange the concept of an arbitrarily large system with that of an infinite system. (This move corresponds to the passage to the "thermodynamic limit.") In such a case it is relatively easy to demonstrate that the results that we obtained for completely isolated systems apply. Although the literature is full of arguments to this effect, let us add another one that seems to us to be particularly simple.

To say that, in the thermodynamic limit, large fluctuations from E are extremely improbable means the following. Let $E(t_1), E(t_2), \ldots$ be

---

[5] For the definition of various concepts of stability, see p. 197.

248

the energy levels of M at the points of time $t_1, t_2, \ldots$ Let $T_{\text{fluc}} = t_{i1}, t_{i2}, \ldots$ be the set of indexes where $E(t_{ij}) \neq E$. Let us recall that a set of indexes $A \subseteq \Pi Z$ has zero density in Z iff $\lim_{n \to \infty} 1/n\{A \cap \{1, 2, \ldots n\}\} = 0$. It is easy to see that, in the thermodynamic limit, the density of T_{fluc} is zero. (Otherwise, if the fluctuations from E occurred with small but nonzero frequency, in the limit, they could cause a serious deviation.) This means that we may, in the spirit of measure theory, simply ignore the set T_{fluc} on the ground that it is a negligible set. But ignoring T_{fluc} means that we ignore the points of time when $E(t_{ij}) \neq E$, which means that we treat the system as if it were a perfectly isolated system that always has the energy E. Note that the argument relies crucially on the fact that we are dealing with systems composed of a great many particles; otherwise, the transition to the thermodynamic limit would be problematic and the entire argument would collapse.

To sum up, then, we may treat many-particle systems that are in a state of equilibrium with respect to their surroundings as if they were completely isolated. That also means that we may, if we so wish, regard such systems as stationary systems and represent them as measure-preserving transformations.

III.4. FORMULATION OF THE FRAMEWORK OF NESM

Is it possible to define SM probabilities in a more general setting? Perhaps a better question to ask is whether we really need to do so. The adherents of the Brussels school believe that, to discuss the problem of irreversibility and other problems of NESM, we need to introduce time-dependent probabilities into the picture. The rest of this appendix will be dedicated to arguing against this point of view. We shall first introduce the approach of Prigogine and his co-workers from the Brussels school and then criticize it.[6]

According to Dougherty's presentation,[7] the Brussels school conception of NESM can be formulated in the following manner. We are given a macroscopic system M and an environment. The usual type of phase space is then constructed for M. (Sometimes, the states of M will be described in quantum mechanics style, that is, as rays in an

---

[6] We shall refer mainly to the formulation of NESM given by Balescu. See ibid, pp. 501–543. Some related ideas are discussed in Prigogine, *From Being to Becoming*, Freeman (New York, 1980), pp. 179–201.

[7] See, for example, Dougherty, *Explaining Statistical Mechanics*, ibid, pp. 851–853.

appropriate Hilbert space.) In addition, claims Dougherty, in NESM we have to specify an appropriately chosen "reduced description." Such a description will include only some of the available information about M. In fact, one is encouraged to select a description that is much less detailed than the "full description," that is, from the specification of the positions and momenta of all of the particles of M. However, the reduced description should not be too impoverished. (If it were too impoverished, the statistical character of the problem would disappear.) For example, it should be more detailed than simply specifying the energy level of M. Once we pick the appropriate reduced descrption, we can define the notion of a reduced trajectory. Instead of describing the evolution of M with the sequence $\{\ldots, w_{-1}w_0w_1w_2, \ldots\}$ of "full states," we can now assign the system the sequence $\{\ldots, w^*_{-1}w^*_0w^*_1w^*_2, \ldots\}$ of the reduced states (w^*_i is obtained from w_i by restricting our attention to the reduced set of parameters.) This mode of description, though, will be quite pointless if, to calculate w^*_{i+1} from w^*_i, we would have to convert w^*_i back to w_i, calculate w_{i+1} by using the Hamiltonian, and then "reduce back" to w^*_{i+1}. The question, then, is whether we can find a way of calculating w^*_{i+1} directly from w^*_i, that is, whether we can find a "reduced Hamiltonian" that would allow us to discard the irrelevant information altogether. According to Dougherty, identifying such a "reduced Hamiltonian" H^* is one of the primary aims of NESM.

A very simple argument, though, will suffice to convince the reader that the search for the reduced Hamiltonian cannot be an easy task. To see this, let us take the following simple example. Let $\mathbf{w} = w_1, w_2$ be a vector, and let O be a linear operator that is given the matrix representation

$$O = \begin{Bmatrix} a_{11} & a_{12} \\ a_{21} & a_{22} \end{Bmatrix};$$

the operation $O\mathbf{w}$ is then

$$\begin{Bmatrix} a_{11} & a_{12} \\ a_{21} & a_{22} \end{Bmatrix} \begin{Bmatrix} w_1 \\ w_2 \end{Bmatrix} = \begin{Bmatrix} a_{11}w_1 + a_{12}w_2 \\ a_{21}w_1 + a_{22}w_2 \end{Bmatrix}.$$

Suppose that we restrict our attention only to the first component w_1. Still, to calculate $O w_i$ we have to know the full description because $O w_1 = a_{11} w_1 + a_{12} w_2$. So, unless a_{12} happens to be zero, we need the full description at $t = 0$ even to calculate the first component at $t = 1$.

Let us say a few words about the mathematical representation of the reduced description when M is given a Hilbert space representation. In such a case, we are looking for a way of decomposing the Hilbert space H and presenting it as $H_1 + H_2$, where H_1 and H_2 are orthogonal subspaces that together span H. (This means that if $w = \{w_1, w_2, \ldots\}$ is a basis of H_1 and $w' = \{w'_1, w'_2, \ldots\}$ is a basis of H_2, then $w \cap w'$ is a basis for H.) The idea is that H_1 contains the relevant information about M while H_2 contains merely the "noise." Once we find such a decomposition of H we can regard the members of H_1 as reduced descriptions of the states of H. Now, let U_t be the operator that assigns for each state $s \in H$ its future at t. Once we are given a decomposition of H, the next step is to look for a way of decomposing U_t as well. That is, we are looking for a way of representing U_t as a combination of U'_t and U''_t. The first part, $U'_t : H_1 \to H_1$, is the "relevant" part of U_t that assigns to each reduced description s_0 another reduced description s_t that may be interpreted as the future reduced description at t. The second part, $U''_t : H_2 \to H_2$, is the irrelevant part that hopefully decays as $t \to \infty$. This means that after an appropriate relaxation period the reduced description will be a good approximation, in the sense of the metric of the Hilbert space, of the states of H. Let Π be a projection from H to H_1. Another way of formulating the objective of the decomposition of U_t is that the following relations will hold:

1. Let $s \in H$ be a state of M. For every t, $\Pi(U_t(s)) = U'_t \Pi(s)$.
2. $U''_t(s)$ vanishes as $t \to \infty$.

This formulation of the basic problematics of NESM is rather attractive. Other formulations of NESM rely on the notion of "coarse graining," which we introduced in Chapter 1.[8] This older formulation of the notion of reduced description seems to be an epistemic notion. (The idea of cutting the state space into squares does not have any natural physical interpretation in classical physics.) In contrast, when we think of the choice of a reduced description as a decomposition of H into H_1 and H_2, we obtain a simpler and more natural idea. When we concentrate on the reduced definition we project each vector onto the subspace H_1. The notion of projection, of course, is completely natural from a physical point of view.

---

[8] See Section 1.5.

251

When M is given a phase-space representation the search for the reduced description, as we have seen, is formulated in a slightly different way, mainly because the transition from a classical phase space to a quantum Hilbert space necessitates the introduction of some new notations. The translation, however, is always relatively simple. Instead of the decomposition of H into H_1 and H_2, we are looking for a way of writing the phase space S as a product of two subspaces, S_1 and S_2. This means that we first must present the set of states S as a union of two sets S_1 and S_2 whose intersection is empty. Next we construct the appropriate σ-algebras B_1 and B_2, which are obtained by taking the intersection of all of the members of B with S_1 and S_2, respectively. Finally, we are looking for a way of expressing T in terms of two transformations, $T_1 : S_1 \to S_1$ and $T_2 : S_2 \to S_2$, which are B_1- and B_2-measurable, respectively. The idea is that $T1$ and $T2$ are the "reducts" of T to the respective subspaces. This means, again, that if Π is the projection of S onto S_1, for example, then $\Pi(T(s)) = T1(\Pi(s))$ for every $s \in S$. Note that if we were to find such a decomposition, we could use the measure on $S1$ as our relevant measure, which means that we could set $m(A) = m_1(\Pi(A))$, where m_1 is a measure that is constructed for the reduced description of M. The choice of m_1 would then reflect the idea that the probabilities should correspond to the relevant level of description of the events.

Note that, on the face of it, the formulation of NESM that we are discussing seems to apply to systems that are neither isolated nor in a state of equilibrium with their environment. That is why the conception of the Brussels school seems to involve a genuine expansion of the scope of statistical mechanics. Therefore, it does not seem appropriate to assume that in statistical mechanics we are dealing only with systems whose probabilities are time-invariant. It seems that, at least theoretically, we can be interested in finding an appropriate reduced description of dissipative systems, for example. Dougherty goes even further and proposes taking time-dependent probabilities as our basic notion. This means that we should not introduce the dependency on time only when M is known to be a dissipative system, but also when the disruption of the equilibrium is due to an internal fluctuation. This suggestion is provocative but interesting. However, as we shall see later, it is plagued by internal problems that disqualify it, at least for the time being, from being considered seriously.

252

What then are the conditions on M that guarantee that a reduced description is possible? How easy it is to prove that specific systems of interest satisfy these conditions? What makes the reduced description appropriate? These are the questions with which the founders of the Brussels school are preoccupied. Indeed, these are the issues on the basis of which the merits of the Brussels school should be evaluated. The reduced level of description is identified, the right reduced law of motion is discussed, and comparisons are made to other more traditional approaches to the same topics. We will not attempt to summarize the discussion on all of these important issues. The reader is invited to read the works of Prigogine, Mizra, Balescu, and other participants of the Brussels school. In addition, the reader may consult the above-mentioned papers of Dougherty. In the rest of this appendix we shall dedicate our efforts to understanding the most fundamental aspects of the proposal and then criticize it. To gain a basic understanding we shall attempt to achieve three humble aims: (i) We shall first attempt to understand the physical basis for the view that reduced description is possible. (ii) Next we shall explain the reasons why it is rational to give up some of the information that is available about M. (iii) Finally, we shall inquire about a more mathematical topic, namely, when can we present a system M as an uncoupled product of two of its subsystems. Having accomplished these three aims, we shall turn to our critical remarks.

(i) Let us begin with the first question, namely, the physical basis for the view that reduced descriptions are possible. Suppose that we concentrate on a set of states A that is defined by using parameters that are measurable and consistent with the reduced description. We are interested in calculating the probabilities that the system will assume states that are in a set B. (Again, B is defined by using parameters that are measurable and consistent with the reduced description.) Now, because we are concentrating on A-states, it is natural to formulate the problem as that of finding the transition probability from A-states to B-states. However, the reason why we are interested in A-states is not because such states fulfill certain conditions that are known to hold. On the contrary, the reason for the interest is that the A-state is described by using only relevant parameters; the irrelevant ones have been discarded to obtain the reduced description. Therefore, the question arises as to whether we introduced bias into our calculations by concentrating only on A-states. Indeed, how can we justify the idea of concentrating only on A-states?

As the reader might recall, if M is a measure-preserving system, we may employ the vocabulary of ergodic theory to formulate precisely what type of system M has to be to free us from worries about statistical biases. More generally, we formulated a condition whose satisfaction will guarantee that the choice of reference classes will not present a problem. Recall that such systems are called mixing. Let us review this concept once again. Later, we shall inquire whether the concept is applicable in the context of NESM.

Let us recall the definition of mixing. $\langle S, B, m, T \rangle$ is mixing iff, for every $A, A' \in B$,

$$\lim_{n \to \infty} m(A \cap T^n(A')) = m(A) m(A'). \tag{*}$$

This definition suggests a procedure for measuring the probability $m(A)$ of an arbitrary measurable set A. First let us observe that (*) implies that

$$\lim_{n \to \infty} m(A \cap T^n(A'))/m(A') = m(A).$$

Because $m(A')$ is a constant, we can set

$$\lim_{n \to \infty} [m(A \cap T^n(A'))/m(A')] = m(A).$$

Furthermore, because when $\langle S, B, m, T \rangle$ is a measure-preserving transformation

$$m(A) = m(T^n(A)),$$

we have

$$\lim_{n \to \infty} [m(A \cap T^n(A'))/m(T^n(A'))] = m(A). \tag{**}$$

Now, (**) has an appealing interpretation as a procedure for measuring A. *We measure A by calculating the limit of the average measure of the intersection between A and a test set and its iterates* $A', T(A'), T^2(A'),$ *.... At each case, the measure of the intersection is weighted by the measure of the test set.* Another way of stating the same idea requires the notion of conditional probability:

$$m(A/A') = m(A \cap A')/m(A).$$

254

By using this definition we can rewrite (**) and obtain

$$\lim_{n \to \infty} [m(A/T^n(A'))] = m(A). \qquad (***)$$

The idea behind (***) is that *we measure A by calculating* $lim_{n \to \infty}$ *[m(A/T^n(A')]*. To sum up, then, if M is mixing, then we may use the above method to measure the probability of an arbitrary measurable subset A without worrying about the possibility that we introduced a bias into our calculations by concentrating only on A'-states. In particular, we may employ (**) to calculate the $m(A)$ by using as our "test set" the set of reduced states. If the system is mixing, such a method will yield correct results.

Observe that the paragraph also sheds new light on the other two questions.

(ii) Regarding the rationality of discarding available information, the argument is the following. It is not irrrational to discard available information when mixing systems are concerned because the discarded information will become asymptotically irrelevant in the limit, that is, as $n \to \infty$.

(iii) Let us now turn to the third question, which concerned the special traits of the reduced description. As we stated before, when mixing systems are concerned, every reduced description will have property (***). Hence, in particular, it will apply to the specific conception of the reduced description that was chosen in this context. This reasoning also demonstrates why it is somewhat misleading to say that the reduced description that the physicists of the Brussels school labored to construct has some special properties. Any reduced description would do instead if mixing is required.

In reality, it is very difficult to prove that measure-preserving systems whose origins are stationary Hamiltonian systems are mixing. In fact, the only physically motivated cases where systems are proven to be mixing are Sinai's billiard-ball systems that we discussed in Chapter 2.[9] This point might cast a doubt on the whole enterprise of applying ergodic theory to statistical mechanics. However, the advantage of ergodic theory is, as usual, the interpretation it gives to the probabilities, namely, as weighted relative frequencies.

Let us now return to an issue that has been left open: The notion of mixing is formulated for measure-preserving transformations. Such

[9] See pp. 153–156.

systems are the mathematical representation of stationary systems, which presents a problem: If M is a nonstationary system, what does it mean for it to be mixing? If we cannot find an answer to the question, we will not be able to carry out the Brussels program for NESM because the condition for the possibility of a reduced description involves mixing.

Let us look again at the formula

$$\lim_{n\to\infty}[m(A \cap T^n(A'))/m(T^n(A'))] = m(A). \qquad (**)$$

As we remarked in the second chapter,[10] the basic idea behind the method for measuring the probability of A' will not be affected even if the sets $A',T(A'),\ldots,T^n(A'),\ldots$ do not have the same measure. The quantity

$$m(A \cap A')/m(A')$$

already takes into account the possibility that S is not a stationary system. (The measure of the intersetion is weighted by the measure of the test set.) Hence, even if the measure of the test sets $A',T(A'),\ldots,$ $T^n(A'),\ldots$ is different, the formula

$$\lim_{n\to\infty}[m(A \cap T^n(A'))/m(T^n(A'))] = m(A) \qquad (*)$$

seems to make good sense. One set of considerations, though, has to be addressed. If the quantities $T^n(A')$ become zero, then the expression

$$m(A \cap T^n(A'))/m(T^n(A'))$$

is ill defined. In such a case, the proportion of the A-states among the $T^n(A')$ will not be taken into consideration, and, consequently, our calculations may be mistaken. Therefore, to use procedure (***), we must require at the outset that, for every $A \in B$, if $m(A) \neq 0$, then, for every n, $T^n(A) \neq 0$. Let us call such systems *quasi-stationary*. The conclusion we wish to draw from the preceding argument is that the notion of mixing can be extended without difficulty to systems that are quasi-stationary. Therefore, such systems can provide a framework for the discussion on NESM.

---

[10] See pp. 131–132.

Can the notion of mixing be formulated for other types of systems? One idea presents itself immediately. We may certainly define *weakly quasi-stationary systems* represented as *weakly nonsingular measurable transformations* by requiring only that, for every $A \in B$, when $m(A) \neq 0$, then the set of indexes $0(A) \subset N, 0(A) = \{i \in N : T^i(A) = 0\}$ has density zero. Can we go any further? It is quite doubtful. When $m(T^n(A'))$ equals zero, we simply cannot use the standard measure-theoretic methods to determine the conditional probability $m(A/T^n(A'))$. What is required is "renormalization," namely, a method that would enable us to assign conditional probabilities of the form $m(A/T^n(A'))$ even if $m(T^n(A')) = 0$. Such assignments can be made, but only in a sporadic manner; the discussion on the general principles of such an undertaking is not very well developed. Therefore, if we wish to avoid the necessity of assessing such probabilities, we must require quasi-stationarity or a weakened version thereof.

III.6. A CONCLUSION

Let us summarize the conclusions we wish to draw in this appendix. The members of the Brussels school proposed an attractive formulation of NESM. First, they proposed the idea that in NESM the most important issue is the search for a reduced level of description of the systems being considered; the essence of the reduced description is to get rid of irrelevant information. This reduced level, according to the Brussels school, represents a perfectly objective feature of the world; it is not merely an epistemic feature introduced into the picture by the observer. Mathematically, we represent the reduced description as a projection onto a subspace. The projection brings irreversibility into the picture. Irreversible phenomena express the fact that once we get rid of the useless information we cannot retrieve it. Another issue we must face is that the reduced description might not be informative enough to allow us to calculate the future states of the system given its present reduced state. At this point mixing is introduced; mixing is a necessary condition whose satisfaction will guarantee that we can calculate the future states of the system on the basis of their reduced states. However, mixing is a concept that primarily applies to stationary systems. Therefore, even if NESM was intended to cover nonstationary systems, the reliance on the mixing requirement prohibits it from doing so. Now the strict stationarity requirement can be weakened. We showed how to define quasi-stationarity and weak quasi-stationarity, and argued that even in these cases the notion of mixing still makes sense. However, we believe that even these weaker versions of stationarity do not allow us to discuss systems whose transition probabilities are genuinely time-dependent. Therefore, we do not agree with the assessment that NESM (or at least the Brussels school version thereof) deals successfully with nonstationary systems. Consequently, we also disagree with the view that we must transcend the ergodic approach to SM probabilities. The ergodic approach allows us to define probabilities as weighted averages of relative frequencies. This approach does not allow us to discuss genuinely time-dependent probabilities. Hence, some writers thought it was discredited. However, since mixing is required and, hence, no genuine time-dependence is introduced into NESM, we believe that the argument against the ergodic approach is a moote point. Those cases that require a genuine departure from the framework of the ergodic approach are those for which no satisfactory definition of mixing exists. Hence, we are under no real pressure to depart from the ergodic approach.

Let us end by explaining why weak quasi-stationarity does not usher in genuine time-dependence. It is well known[11] that if T is a continuous transformation on a compact metric space and μ is a stationary measure, then μ is mixing iff, for every measure m (which is not necessarily stationary), the following condition holds: If $m \ll \mu$, then, for every $A \in B$, $\lim_{n \to \infty} m(T^n(A)) = \mu(A)$. If we require that m be quasi-stationary, it is obvious that $m \ll \mu$. Hence, m converges to μ and, as a result, there is no infinite subseries of $T^n(A)$ that converges to another limit. Thus we may regard m as a "notational variant" of μ rather than a genuinely different measure. The restriction of the theorem to transformations on compact metric spaces does not limit the physical applicability of the theorem.

NOTE ADDED IN PROOF

The third appendix was somewhat of an afterthought, and we decided to include it with some hesitation. It does not do justice to the large and complex literature on NESM, and that was not the reason for its inclusion. It should be read mostly as an indication of some problem in the foundations of the concept of probability in the context of NESM.

---

[11] See Walters, op cit., theorem 6.12.

Author Index

Subject Index

262

dispositions, 110–111, 115, 226
de dicto, 166
de re, 166
decay of autocorrelation, 91, 213
decision theory, 228
degeneration, 66
degrees of belief, 27, 35, 133, 162, 192, 238
dense sets, 144, 156, 173, 175
descriptive set theory, 147
differential eqations, 163, 165
diffusion, 67
dilluted gas, 46
dualism, 230
duality between measure and category, 184
Dutch book arguments, 233
dynamic approach, 35, 52
dynamical systems, 76, 78, 112
dynamics of belief, 193

eliminative approaches, 152, 235–236
to probability, 36, 134–150, 234
emergence of properties, 121–131
energy conservation law, 16–17, 48
ensembles, 22, 26, 30–33, 52, 54
canonical, 22, 43–44, 49–51, 52, 53, 123, 208
equality of microcanonical and canonical ensembles, 53
grand canonical, 23, 123
microcanonical, 23, 53, 92
subensembes of canonical ensembles, 53
entropy, 42, 45, 48, 51, 54–55, 65
maximum entropy principle, 30, 41–45, 50, 55–60, 72
Erdös–Sierpinski theorem, 154, 158
ergodic approach, 26, 29, 33, 36, 55, 60, 70, 73, 87–96, 104
ergodic decomposition theorem, 105, 203–204
ergodic hypothesis, 72, 135, 171
Boltzmann on, 141
Maxwell on, 147–149, 172
proof of the impossibility of, 73, 87

topological vs. probabilistic versions of, 171–177
ergodic measures, 105
unique ergodicity, 206
ergodic systems, 59–60, 75, 77–78, 81, 84, 89, 104, 113–115, 161–163, 203–204
ergodic theorems, 55, 73–80, 155
Birkhof, 77, 84
coarse grain vs. fine grain, 83–84
Khinchin, 119
von Neuman, 151–152
quantum, 120
ergodic theory, 5, 6, 73–80, 83–87, 155
descriptive, 201–207
equality of time and space averages, 90
equilibrium, 6, 16, 22, 37–41, 43, 51, 52, 53, 75, 78, 92, 133, 154
equipartition theorem, 20, 68
equivalence by decomposition, 146
equivalent volumes, law of, 17
errors, theory of, 10–12, 14, 21–22, 24, 27, 209
exchangeability, 104, 106, 107, 111–112
expectation, 96, 106
of time averages, 105
explanation, 240–244
statistical, 240–244

ferromagnetism, 121–131
finitism, 68, 100, 102, 136, 150
fluctuations, 16, 248
frequentist approach to probabilities, 75, 80, 81, 82, 84, 88, 102–104, 113–114, 226, 238, 239

games of chance, 224–225
generalized averages, *see* averages, generalized
generic property, 156
Gibbs's paradox, 23, 33
grand canonical distributions, *see* ensembles, grand canonical

263

reduction, 123, 124, 151
  of thermodymamics to mechanics,
    69, 72, 120, 121, 124, 128, 133,
    150
reference class problem, 169, 170,
    242–244
regional transitivity, 171–172, 177–
    181
regular measures, 157
reversed motion, 16, 17
relaxation time, 85–86, 127, 248
Ryll–Nardzewski theorem, 111

second law of thermodynamics, 4, 26,
    51, 65–67, 72, 116, 118, 119
sensitivity to initial conditions, 127,
    126, 130, 133, 174
seperation axioms, 98, 219
sets
  of first category, *see* category
  of second category, *see* category
σ-additivity, 217, 221, 234
σ-algebra, 217
σ-ideal, 158, 161, 218
similarity, 170
  subjectivist account of, 194
shift transformation, 79, 100–101
singularities, 133
  and phase transitions, 126–127
small sets, 158, 167–168
  sets of first category, 158
  sets of measure zero, 158
social science, 100
space-filling curves, 140
specific heats, 20, 25, 30, 32, 71
spin, 122, 123, 124
spin echo effect, 38
spontaneous magnetism, 123, 124, 130
square-integrable functions, 77
stability, 88, 132–133, 164, 177–181,
    198–199
  instability, 132
  Lagrange stability, 164
  Lyapunov stability, 164
  Poisson stability, 164
  stable sets, 19
  stable theories, 169, 182

stability dogma, 169–170
structural stability, 187
stationarity, 101, 107–115, 119, 246
statistical inference, 28, 29, 34, 53
stochastic hierarchy, 94
stochastic processes, 79, 93
stosszahlansatz, 18, 24, 71, 90, 93, 95
subjectivism, 229, 231
  permissive, 104, 107
  subjective, 27, 36, 38, 51, 55, 76, 79,
    96–107, 107–115, 192–200,
    232–234
  ultrasubjectivism, 28, 29, 38, 41, 44,
    45, 55, 234–237
sum functions, 90, 91
symmetry, 226–227, 230
  breaking, 127
  matrixes, 16
  time, 3, 16
symplectic structure, 105

tenacity requirement, 109
theory of errors, *see* error theory
thermodynamics, 53, 118–119
  quantum thermodynamics, 120
  thermodynamic limit, 25, 53, 93,
    126, 248–249
time average, *see* averages
time invariance, 99, 108
time symmetry, *see* symmetry
topology, 156–166, 218–221
  topological entropy, 183
  topological groups, 35, 74
  topological reconstruction of
    statistical mechanics, 171–177
  topological space, 218
topological dynamics, 155, 159–171,
    171–177
transition probabilities, 246
transitive systems, 143, 165
transport equation, 47–49

ultrasubjectivism, *see* subjectivism
Umkehreinwand, 72
uncertainty relations, 84

velocity distribution, 22

Printed in the United States
By Bookmasters